CCNP BCMSN
Portable Command Guide

Scott Empson

Cisco Press

800 East 96th Street
Indianapolis, IN 46240 USA

CCNP BCMSN Portable Command Guide

Scott Empson

Copyright © 2007 Cisco Systems, Inc.

Published by:
Cisco Press
800 East 96th Street
Indianapolis, IN 46240 USA

ISBN-10: 1-58720-188-7

ISBN-13: 978-1-58720-188-2

Printed in the United States of America 1 2 3 4 5 6 7 8 9 0

First Printing   June 2007

Library of Congress Cataloging-in-Publication Data

Empson, Scott.

  CCNP BCMSN portable command guide / Scott Empson.

     p. cm.

  ISBN 978-1-58720-188-2 (pbk.)

  1. Computer networks--Problems, exercises, etc. 2. Computer networks--Examinations--Study guides. 3. Packet switching (Data transmission)--Examinations--Study guides. I. Title.

  TK5105.8.C57E57 2007

  004.6'6--dc22

                                       2007019367

Warning and Disclaimer

This book is designed to provide information about the Certified Cisco Networking Professional (CCNP) 642-812 Building Cisco Multilayer Switched Networks (BCMSN) exam and the commands needed at this level of network administration. Every effort has been made to make this book as complete and as accurate as possible, but no warranty or fitness is implied.

The information is provided on an "as is" basis. The author, Cisco Press, and Cisco Systems, Inc. shall have neither liability nor responsibility to any person or entity with respect to any loss or damages arising from the information contained in this book or from the use of the discs or programs that may accompany it.

The opinions expressed in this book belong to the author and are not necessarily those of Cisco Systems, Inc.

Trademark Acknowledgments

All terms mentioned in this book that are known to be trademarks or service marks have been appropriately capitalized. Cisco Press or Cisco Systems, Inc. cannot attest to the accuracy of this information. Use of a term in this book should not be regarded as affecting the validity of any trademark or service mark.

Feedback Information

At Cisco Press, our goal is to create in-depth technical books of the highest quality and value. Each book is crafted with care and precision, undergoing rigorous development that involves the unique expertise of members from the professional technical community.

Readers' feedback is a natural continuation of this process. If you have any comments regarding how we could improve the quality of this book, or otherwise alter it to better suit your needs, you can contact us through email at feedback@ciscopress.com. Please make sure to include the book title and ISBN in your message.

We greatly appreciate your assistance.

Corporate and Government Sales

Cisco Press offers excellent discounts on this book when ordered in quantity for bulk purchases or special sales.

For more information please contact: U.S. Corporate and Government Sales 1-800-382-3419 corpsales@pearsontechgroup.com

For sales outside the U.S. please contact: International Sales international@pearsoned.com

Publisher: Paul Boger

Associate Publisher: David Dusthimer

Executive Editor: Mary Beth Ray

Cisco Representative: Anthony Wolfenden

Cisco Press Program Manager: Jeff Brady

Managing Editor: Patrick Kanouse

Senior Development Editor: Christopher Cleveland

Project Editor: Seth Kerney

Copy Editor: Keith Cline

Proofreader: Water Crest Publishing, Inc.

Technical Editors: Tami Day-Orsatti and David Kotfila

Team Coordinator: Vanessa Evans

Book Designer: Louisa Adair

Composition: Mark Shirar

CISCO.

Americas Headquarters	Asia Pacific Headquarters	Europe Headquarters
Cisco Systems, Inc.	Cisco Systems, Inc.	Cisco Systems International BV
170 West Tasman Drive	168 Robinson Road	Haarlerbergpark
San Jose, CA 95134-1706	#28-01 Capital Tower	Haarlerbergweg 13-19
USA	Singapore 068912	1101 CH Amsterdam
www.cisco.com	www.cisco.com	The Netherlands
Tel: 408 526-4000	Tel: +65 6317 7777	www-europe.cisco.com
800 553-NETS (6387)	Fax: +65 6317 7799	Tel: +31 0 800 020 0791
Fax: 408 527-0883		Fax: +31 0 20 357 1100

Cisco has more than 200 offices worldwide. Addresses, phone numbers, and fax numbers are listed on the Cisco Website at **www.cisco.com/go/offices.**

©2007 Cisco Systems, Inc. All rights reserved. CCVP, the Cisco logo, and the Cisco Square Bridge logo are trademarks of Cisco Systems, Inc.; Changing the Way We Work, Live, Play, and Learn is a service mark of Cisco Systems, Inc.; and Access Registrar, Aironet, BPX, Catalyst, CCDA, CCDP, CCIE, CCIP, CCNA, CCNP, CCSP, Cisco, the Cisco Certified Internetwork Expert logo, Cisco IOS, Cisco Press, Cisco Systems, Cisco Systems Capital, the Cisco Systems logo, Cisco Unity, Enterprise/Solver, EtherChannel, EtherFast, EtherSwitch, Fast Step, Follow Me Browsing, FormShare, GigaDrive, GigaStack, HomeLink, Internet Quotient, IOS, IP/TV, iQ Expertise, the iQ logo, iQ Net Readiness Scorecard, iQuick Study, LightStream, Linksys, MeetingPlace, MGX, Networking Academy, Network Registrar, Packet, PIX, ProConnect, RateMUX, ScriptShare, SlideCast, SMARTnet, StackWise, The Fastest Way to Increase Your Internet Quotient, and TransPath are registered trademarks of Cisco Systems, Inc. and/or its affiliates in the United States and certain other countries.

All other trademarks mentioned in this document or Website are the property of their respective owners. The use of the word partner does not imply a partnership relationship between Cisco and any other company. (0609R)

About the Author

Scott Empson is currently the assistant program chair of the bachelor of applied information systems technology degree program at the Northern Alberta Institute of Technology in Edmonton, Alberta, Canada, where he teaches Cisco routing, switching, and network design courses in a variety of different programs—certificate, diploma, and applied degree—at the post-secondary level. Scott is also the program coordinator of the Cisco Networking Academy Program at NAIT, a Regional Academy covering central and northern Alberta. He has earned three undergraduate degrees: a bachelor of arts, with a major in English; a bachelor of education, again with a major in English/language arts; and a bachelor of applied information systems technology, with a major in network management. He currently holds several industry certifications, including CCNP, CCDA, CCAI, and Network+. Before instructing at NAIT, he was a junior/senior high school English/language arts/computer science teacher at different schools throughout northern Alberta. Scott lives in Edmonton, Alberta, with his wife, Trina, and two children, Zachariah and Shaelyn, where he enjoys reading, performing music on the weekend with his classic rock band "Miss Understood," and studying the martial art of Taekwon-Do.

About the Technical Reviewers

Tami Day-Orsatti (CCSI, CCDP, CCNP, CISSP, MCT, MCSE 2000/2003: Security) is an IT networking and security instructor for T^2 IT Training. She is responsible for the delivery of authorized Cisco, $(ISC)^2$, and Microsoft classes. She has more than 23 years in the IT industry working with many different types of organizations (private business, city and federal government, and the Department of Defense), providing project management and senior-level network and security technical skills in the design and implementation of complex computing environments.

David Kotfila (CCNP, CCAI) is the director of the Cisco Academy at Rensselaer Polytechnic Institute (RPI), Troy, New York. Under his direction, more than 125 students have received their CCNP, and 6 students have obtained their CCIE. David is a consultant for Cisco, working as a member of the CCNP assessment group. His team at RPI is authoring the four new CCNP lab books for the Academy program. David has served on the National Advisory Council for the Academy program for four years. Previously, he was the senior training manager at PSINet, a Tier 1 global Internet service provider. When David is not staring at his beautiful wife, Kate, or talking with his two wonderful children, Chris and Charis, he likes to kayak and lift weights.

Dedications

This book is dedicated to Trina, Zach, and Shae, without whom I couldn't have made it through those long nights of editing.

Acknowledgments

Anyone who has ever has anything to do with the publishing industry knows that it takes many, many people to create a book. It may be my name on the cover, but there is no way that I can take credit for all that occurred to get this book from idea to publication. Therefore, I must thank a number of people.

The team at Cisco Press—once again, you amaze me with your professionalism and the ability to make me look good. Mary Beth, Chris, Patrick, and Seth—thank you for your continued support and belief in my little engineering journal.

To my technical reviewers, Tami and David—thanks for keeping me on track and making sure that what I wrote was correct and relevant.

To the staff of the Cisco office here in Edmonton—thanks for putting up with me and my continued requests to borrow equipment for development and validation of the concepts in this book.

To Rick Graziani—thank you for showing me how to present this material to my students in a fun and entertaining way, and in an educational manner.

Finally, big thanks go out to Hans Roth. There are not enough superlatives in the dictionary to describe Hans and his dedication to not only education, but also to the world of networking in general. While I was working on this series of books, Hans decided that he needed to leave the Ivory Tower of Education and get his hands dirty again in industry. So what better way to get back into the swing of things than to go to Africa and design and help install a new converged infrastructure for an entire country? He also had enough time to listen to my ideas, make suggestions, and build most of the diagrams that are in this book. His input has always been invaluable, and for that, I thank you.

This Book Is Safari Enabled

The Safari® Enabled icon on the cover of your favorite technology book means the book is available through Safari Bookshelf. When you buy this book, you get free access to the online edition for 45 days.

Safari Bookshelf is an electronic reference library that lets you easily search thousands of technical books, find code samples, download chapters, and access technical information whenever and wherever you need it.

To gain 45-day Safari Enabled access to this book:

- Go to http://www.ciscopress.com/safarienabled
- Complete the brief registration form
- Enter the coupon code MIAJ-VHPC-DMAY-J2BE-99F4

If you have difficulty registering on Safari Bookshelf or accessing the online edition, please e-mail customer-service@safaribooksonline.com.

Contents at a Glance

Introduction   xiii

Chapter 1   Network Design Requirements   1

Chapter 2   VLANs   3

Chapter 3   STP and EtherChannel   17

Chapter 4   Inter-VLAN Routing   43

Chapter 5   High Availability   59

Chapter 6   Wireless Client Access   75

Chapter 7   Minimizing Service Loss and Data Theft   101

Chapter 8   Voice Support in Campus Switches   121

Appendix   Create Your Own Journal Here   125

Contents

Introduction   xiii

Chapter 1   Network Design Requirements   1

Cisco Service-Oriented Network Architecture   1
Cisco Enterprise Composite Network Model   2

Chapter 2   VLANs   3

Creating Static VLANs   3
Using VLAN-Configuration Mode   3
Using VLAN Database Mode   4
Assigning Ports to VLANs   5
Using the range Command   5
Dynamic Trunking Protocol   5
Setting the Encapsulation Type   6
Verifying VLAN Information   7
Saving VLAN Configurations   7
Erasing VLAN Configurations   8
Verifying VLAN Trunking   9
VLAN Trunking Protocol   9
Using Global Configuration Mode   9
Using VLAN Database Mode   10
Verifying VTP   12
Configuration Example: VLANs   13
3560 Switch   13
2960 Switch   15

Chapter 3   STP and EtherChannel   17

Spanning Tree Protocol   18
Enabling Spanning Tree Protocol   18
Configuring the Root Switch   18
Configuring a Secondary Root Switch   19
Configuring Port Priority   19
Configuring the Path Cost   20
Configuring the Switch Priority of a VLAN   20
Configuring STP Timers   21
Verifying STP   21
Optional STP Configurations   22
PortFast   22
BPDU Guard   22
BPDU Filtering   23
UplinkFast   24

BackboneFast   24
Root Guard   24
Loop Guard   25
Unidirectional Link Detection   25
Changing the Spanning-Tree Mode   26
Extended System ID   27
Enabling Rapid Spanning Tree   27
Enabling Multiple Spanning Tree   28
Verifying MST   29
Troubleshooting Spanning Tree   29
Configuration Example: STP   30
Core Switch (3560)   30
Distribution 1 Switch (3560)   31
Distribution 2 Switch (3560)   32
Access 1 Switch (2960)   33
Access 2 Switch (2960)   34
EtherChannel   34
Interface Modes in EtherChannel   35
Guidelines for Configuring EtherChannel   35
Configuring L2 EtherChannel   36
Configuring L3 EtherChannel   36
Verifying EtherChannel   37
Configuration Example: EtherChannel   38
DLSwitch (3560)   39
ALSwitch1 (2960)   40
ALSwitch2 (2960)   41

Chapter 4   Inter-VLAN Routing   43
Configuring Cisco Express Forwarding   43
Verifying CEF   44
Troubleshooting CEF   44
Inter-VLAN Communication Using an External Router:
   Router-on-a-Stick   45
Inter-VLAN Communication Tips   46
Inter-VLAN Communication on a Multilayer Switch Through a
   Switch Virtual Interface   46
Removing L2 Switchport Capability of a Switch Port   46
Configuring Inter-VLAN Communication   47
Configuration Example: Inter-VLAN Communication   48
ISP Router   49
CORP Router   50
L2Switch2 (Catalyst 2960)   52

L3Switch1 (Catalyst 3560)   54
L2Switch1 (Catalyst 2960)   56

Chapter 5   High Availability   59
Hot Standby Routing Protocol   59
Configuring HSRP   59
Verifying HSRP   60
HSRP Optimization Options   60
Preempt   60
HSRP Message Timers   61
Interface Tracking   62
Debugging HSRP   62
Virtual Router Redundancy Protocol   62
Configuring VRRP   63
Verifying VRRP   64
Debugging VRRP   64
Gateway Load Balancing Protocol   65
Configuring GLBP   65
Verifying GLBP   68
Debugging GLBP   68
Configuration Example: HSRP   69
Router 1   69
Router 2   70
Configuration Example: GLBP   71
Router 1   72
Router 2   73

Chapter 6   Wireless Client Access   75
Configuration Example: 4402 WLAN Controller Using the
Configuration Wizard   75
Configuration Example: 4402 WLAN Controller Using the
Web Interface   84
Configuration Example: Configuring a 3560 Switch to Support
WLANs and APs   94
Configuration Example: Configuring a Wireless Client   96

Chapter 7   Minimizing Service Loss and Data Theft   101
Configuring Static MAC Addresses   101
Switch Port Security   102
Verifying Switch Port Security   103
Sticky MAC Addresses   104
Mitigating VLAN Hopping: Best Practices   105

Configuring Private VLANs   105
Verifying PVLANs   106
Configuring Protected Ports   107
VLAN Access Maps   107
Verifying VLAN Access Maps   109
Configuration Example: VLAN Access Maps   109
DHCP Snooping   111
Verifying DHCP Snooping   113
Dynamic ARP Inspection   113
Verifying DAI   114
802.1x Port-Based Authentication   114
Cisco Discovery Protocol Security Issues   116
Configuring the Secure Shell Protocol   117
vty ACLs   117
Restricting Web Interface Sessions with ACLs   118
Disabling Unneeded Services   118
Securing End-Device Access Ports   119

Chapter 8   Voice Support in Campus Switches   121
Attaching a Cisco IP Phone   121
Verifying Configuration After Attaching a Cisco IP Phone   123
Configuring AutoQoS: 2960/3560   123
Verifying AutoQoS Information: 2960/3560   124
Configuring AutoQoS: 6500   124
Verifying AutoQoS Information: 6500   124

Appendix   Create Your Own Journal Here   125

Icons Used in This Book

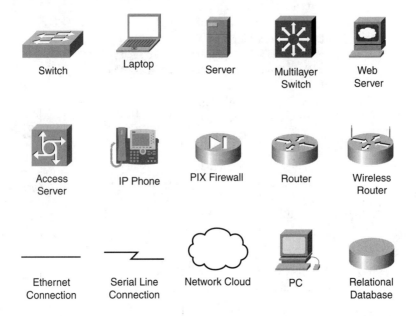

| Switch | Laptop | Server | Multilayer Switch | Web Server |

| Access Server | IP Phone | PIX Firewall | Router | Wireless Router |

| Ethernet Connection | Serial Line Connection | Network Cloud | PC | Relational Database |

Command Syntax Conventions

The conventions used to present command syntax in this book are the same conventions used in the IOS Command Reference. The Command Reference describes these conventions as follows:

- **Boldface** indicates commands and keywords that are entered literally as shown. In actual configuration examples and output (not general command syntax), boldface indicates commands that are manually input by the user (such as a **show** command).
- *Italics* indicate arguments for which you supply actual values.
- Vertical bars (|) separate alternative, mutually exclusive elements.
- Square brackets [ ] indicate optional elements.
- Braces { } indicate a required choice.
- Braces within brackets [{ }] indicate a required choice within an optional element.

Introduction

Welcome to BCMSN! In 2006, Cisco Press came to me and told me, albeit very quietly, that there was going to be a major revision of the CCNP certification exams. They then asked whether I would be interested in working on a command guide in the same fashion as my previous books for Cisco Press: the Cisco Networking Academy Program *CCNA Command Quick Reference* and the *CCNA Portable Command Guide*. The original idea was to create a single-volume command summary for all four of the new CCNP exams. However, early on in my research, I quickly discovered that there was far too much information in the four exams to create a single volume—that would have resulted in a book that was neither portable nor quick as a reference. So, I jokingly suggested that they let me author four books—one for each exam. Well, I guess you have to be careful what you wish for, because Cisco Press readily agreed. They were so excited about the idea that they offered to cut the proposed writing time by a few months to get these books to market faster. How nice of them, don't you think?

This book is the second in a four-volume set that attempts to summarize the commands and concepts that you need to pass one of the CCNP certification exams—in this case, the Building Cisco Multilayer Switched Networks exam. It follows the format of my previous books, which are in fact a cleaned-up version of my own personal engineering journal. I have long been a fan of what I call the "Engineering Journal"—a small notebook that can be carried around and that contains little nuggets of information—commands that you forget, the IP addressing scheme of some remote part of the network, little reminders about how to do something you only have to do once or twice a year, but is vital to the integrity and maintenance of your network. This journal has been a constant companion by my side for the past eight years; I only teach some of these concepts every second or third year, so I constantly need to refresh commands and concepts, and learn new commands and ideas as they are released by Cisco. With the creation of two brand-new CCNP exams, the amount of new information out there is growing on an almost daily basis. There is always a new white paper to read, a new Webinar to view, another slideshow from a Networkers session that I didn't get to. My journals are the best way for me to review because they are written in my own words, words that I can understand. At least, I better understand them, because if I didn't, I have only myself to blame.

To make this guide a more realistic one for you to use, the folks at Cisco Press have decided to continue with my request for an appendix of blank pages—pages that are for you to put your own personal touches—your own configurations, commands that are not in this book but are needed in your world, and so on. That way this book will look less like my journal and more like your own.

I hope that you learn as much from reading this guide as I did when I wrote it.

Networking Devices Used in the Preparation of This Book

To verify the commands in this book, I had to try them out on a few different devices. The following is a list of the equipment I used in the writing of this book:

- C2620 router running Cisco IOS Software Release 12.3(7)T, with a fixed Fast Ethernet interface, a WIC-2A/S serial interface card, and a NM-1E Ethernet interface
- C2811 ISR bundle with PVDM2, CMME, a WIC-2T, FXS and FXO VICs, running 12.4(3g) IOS
- WS-C3560-24-EMI Catalyst switch, running 12.2(25)SE IOS
- WS-C3550-24-EMI Catalyst switch, running 12.1(9)EA1c IOS
- WS-C2960-24TT-L Catalyst switch, running 12.2(25)SE IOS
- WS-C2950-12 Catalyst switch, running Version C2950-C3.0(5.3)WC(1) Enterprise Edition software
- AIR-WLC4402 Wireless LAN Controller

These devices were not running the latest and greatest versions of Cisco IOS Software. Some of it is quite old.

Those of you familiar with Cisco devices will recognize that a majority of these commands work across the entire range of the Cisco product line. These commands are not limited to the platforms and Cisco IOS versions listed. In fact, in most cases, these devices are adequate for someone to continue his or her studies beyond the CCNP level, too.

Who Should Read This Book

This book is for those people preparing for the CCNP BCMSN exam, whether through self-study, on-the-job training and practice, study within the Cisco Academy Program, or study through the use of a Cisco Training Partner. There are also some handy hints and tips along the way to make life a bit easier for you in this endeavor. It is small enough that you will find it easy to carry around with you. Big, heavy textbooks might look impressive on your bookshelf in your office, but can you really carry them all around with you when you are working in some server room or equipment closet somewhere?

Organization of This Book

This book follows the list of objectives for the CCNP BCMSN exam:

- **Chapter 1, "Network Design Requirements"**—Provides an overview of the two different design models from Cisco—the Service-Oriented Network Architecture and the Enterprise Composite Network Model.
- **Chapter 2, "VLANs"**—Describes how to configure, verify, and troubleshoot VLANs, including topics such as Dynamic Trunking Protocol (DTP) and VLAN Trunking Protocol (VTP).

- **Chapter 3, "STP and EtherChanel"**—Describes how to configure, verify, and troubleshoot Spanning Tree Protocol (STP), including topics such as configuring the root switch; port priorities; timers; PortFast; BPDU Guard; UplinkFast and BackboneFast; Configuring L2 and L3 EtherChannel; load balancing; and verifying EtherChannel.

- **Chapter 4, "Inter-VLAN Routing"**—Describes how to configure, verify, and troubleshoot inter-VLAN routing, including topics such as router-on-a-stick; switch virtual interfaces; Cisco Express Forwarding (CEF); and creating a routed port on a switch.

- **Chapter 5, "High Availability"**—Covers topics such as Hot Standby Router Protocol (HSRP), Virtual Router Redundancy Protocol (VRRP), and Gateway Load Balancing Protocol (GLBP).

- **Chapter 6, "Wireless Client Access"**—Describes how to configure and verify the configuration of a wireless LAN controller using both the Command-Line Wizard and the GUI Wizard.

- **Chapter 7, "Minimizing Service Loss and Data Theft"**—Covers topics such as port security, sticky MAC addresses, private VLANs, VLAN access maps, DHCP snooping, dynamic ARP inspection, 802.1x authentication, Cisco Discovery Protocol (CDP) issues, Secure Shell (SSH), vty access control lists (ACL), disabling unneeded services, and securing end device access ports.

- **Chapter 8, "Voice Support in Campus Switches"**—Covers topics such as attaching a Cisco IP Phone, configuring AutoQos on a 2960/3560 switch, configuring AutoQos on a 6500, and verifying AutoQoS information.

Did I Miss Anything?

I am always interested to hear how my students, and now readers of my books, do on both vendor exams and future studies. If you would like to contact me and let me know how this book helped you in your certification goals, please do so. Did I miss anything? Let me know. I can't guarantee I'll answer your e-mail message, but I can guarantee that I will read all of them. My e-mail address is ccnpguide@empson.ca.

CHAPTER 1

Network Design Requirements

This chapter provides information concerning the following topics:

- Cisco Service-Oriented Network Architecture
- Cisco Enterprise Composite Network Model

No commands are associated with this module of the CCNP BCMSN course objectives.

Cisco Service-Oriented Network Architecture

Figure 1-1 shows the Cisco Service-Oriented Network Architecture (SONA) framework.

Figure 1-1   Cisco SONA Framework

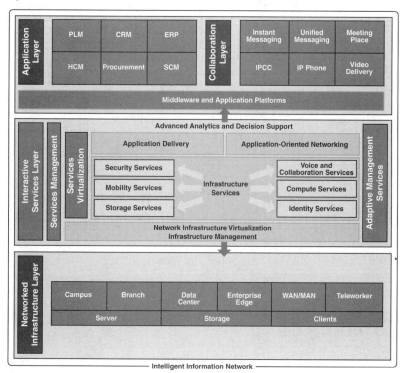

Cisco Enterprise Composite Network Model

Figure 1-2 shows the Cisco Enterprise Composite Network Model.

Figure 1-2   Cisco Enterprise Composite Network Model

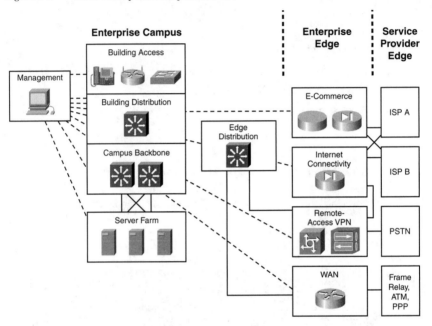

CHAPTER 2

VLANs

This chapter provides information and commands concerning the following topics:

- Creating static VLANs
  - Using VLAN-configuration mode
  - Using VLAN Database mode
- Assigning ports to VLANs
- Using the **range** command
- Dynamic Trunking Protocol (DTP)
- Setting the encapsulation type
- Verifying VLAN information
- Saving VLAN configurations
- Erasing VLAN configurations
- Verifying VLAN trunking
- VLAN Trunking Protocol (VTP)
  - Using VLAN Database mode
  - Using global configuration mode
- Verifying VTP

Creating Static VLANs

Static VLANs occur when a switch port is manually assigned by the network administrator to belong to a VLAN. Each port is associated with a specific VLAN. By default, all ports are originally assigned to VLAN 1. There are two different ways to create VLANs:

- Using the VLAN-configuration mode, which is the recommended method of creating VLANs
- Using the VLAN Database mode (which should not be used, but is still available)

Using VLAN-Configuration Mode

`Switch(config)#vlan 3`	Creates VLAN 3 and enters VLAN-config mode for further definitions
`Switch(config-vlan)#name Engineering`	Assigns a name to the VLAN. The length of the name can be from 1 to 32 characters.

`Switch(config-vlan)#`**`exit`**	Applies changes, increases the revision number by 1, and returns to global configuration mode
`Switch(config)#`	

NOTE: This method is the only way to configure extended-range VLANs (VLAN IDs from 1006–4094).

NOTE: Regardless of the method used to create VLANs, the VTP revision number is increased by one each time a VLAN is created or changed.

Using VLAN Database Mode

CAUTION: The VLAN Database mode has been deprecated and will be removed in some future Cisco IOS release. It is recommended to use only VLAN-configuration mode.

`Switch#`**`vlan database`**	Enters VLAN Database mode
`Switch(vlan)#`**`vlan 4 name Sales`**	Creates VLAN 4 and names it Sales. The length of the name can be from 1 to 32 characters.
`Switch(vlan)#`**`vlan 10`**	Creates VLAN 10 and gives it a name of VLAN0010 as a default
`Switch(vlan)#`**`apply`**	Applies changes to the VLAN database and increases the revision number by 1
`Switch(vlan)#`**`exit`**	Applies changes to the VLAN database, increases the revision number by 1, *and* exits VLAN Database mode
`Switch#`	

NOTE: You must apply the changes to the VLAN database for the changes to take effect. You must use either the **apply** command or the **exit** command to do so. Using the Ctrl-z command to exit out of the VLAN database does not work in this mode because it will abort all changes made to the VLAN database—you must either use **exit** or **apply** and then the **exit** command.

Assigning Ports to VLANs

`Switch(config)#interface fastethernet 0/1`	Moves to interface configuration mode
`Switch(config-if)#switchport mode access`	Sets the port to access mode
`Switch(config-if)#switchport access vlan 10`	Assigns this port to VLAN 10

NOTE: When the **switchport mode access** command is used, the port will operate as a nontrunking, single VLAN interface that transmits and receives nonencapsulated frames.

An access port can belong to only one VLAN.

Using the range Command

`Switch(config)#interface range fastethernet 0/1 - 9`	Enables you to set the same configuration parameters on multiple ports at the same time
	NOTE: There is a space before and after the hyphen in the **interface range** command.
`Switch(config-if-range)#switchport mode access`	Sets ports 1–9 as access ports
`Switch(config-if-range)#switchport access vlan 10`	Assigns ports 1–9 to VLAN 10

Dynamic Trunking Protocol

`Switch(config)#interface fastethernet 0/1`	Moves to interface configuration mode
`Switch(config-if)#switchport mode dynamic desirable`	Makes the interface actively attempt to convert the link to a trunk link
	NOTE: With the **switchport mode dynamic desirable** command set, the interface will become a trunk link if the neighboring interface is set to **trunk**, **desirable**, or **auto**.

`Switch(config-if)#`**`switchport mode dynamic auto`**	Makes the interface able to convert into a trunk link
	NOTE: With the **switchport mode dynamic auto** command set, the interface will become a trunk link if the neighboring interface is set to **trunk** or **desirable**.
`Switch(config-if)#`**`switchport nonegotiate`**	Prevents the interface from generating DTP frames.
	NOTE: Use the **switchport mode nonegotiate** command only when the interface switchport mode is **access** or **trunk**. You must manually configure the neighboring interface to establish a trunk link.
`Switch(config-if)#`**`switchport mode trunk`**	Puts the interface into permanent trunking mode and negotiates to convert the link into a trunk link
	NOTE: With the **switchport mode trunk** command set, the interface becomes a trunk link even if the neighboring interface is not a trunk link.

TIP: The default mode is dependent on the platform. For the 2960 and 3560, the default mode is dynamic auto.

Setting the Encapsulation Type

`3560Switch(config)#`**`interface fastethernet 0/1`**	Moves to interface config mode
`3560Switch(config-if)#`**`switchport mode trunk`**	Puts the interface into permanent trunking mode and negotiates to convert the link into a trunk link
`3560Switch(config-if)#`**`switchport trunk encapsulation isl`**	Specifies Inter-Switch Link (ISL) encapsulation on the trunk link

`3560Switch(config-if)#switchport trunk encapsulation dot1q`	Specifies 802.1Q encapsulation on the trunk link
`3560Switch(config-if)#switchport trunk encapsulation negotiate`	Specifies that the interface negotiate with the neighboring interface to become either an ISL or Dot1Q trunk, depending on the capabilities or configuration of the neighboring interface

TIP: With the **switchport trunk encapsulation negotiate** command set, the preferred trunking method is ISL.

CAUTION: The 2960 series switch supports only Dot1Q trunking.

Verifying VLAN Information

`Switch#show vlan`	Displays VLAN information
`Switch#show vlan brief`	Displays VLAN information in brief
`Switch#show vlan id 2`	Displays information of VLAN 2 only
`Switch#show vlan name marketing`	Displays information of VLAN named marketing only
`Switch#show interfaces vlan x`	Displays interface characteristics for the specified VLAN

Saving VLAN Configurations

The configurations of VLANs 1 through 1005 are always saved in the VLAN database. As long as the **apply** or the **exit** command is executed in VLAN Database mode, changes are saved. If you are using VLAN-configuration mode, using the **exit** command will also save the changes to the VLAN database.

If the VLAN database configuration is used at startup, and the startup configuration file contains extended-range VLAN configuration, this information is lost when the system boots.

If you are using VTP transparent mode, the configurations are also saved in the running configuration, and can be saved to the startup configuration using the **copy running-config startup-config** command.

If the VTP mode is transparent in the startup configuration, and the VLAN database and the VTP domain name from the VLAN database matches that in the startup configuration file, the VLAN database is ignored (cleared), and the VTP and VLAN configurations in the startup configuration file are used. The VLAN database revision number remains unchanged in the VLAN database.

Erasing VLAN Configurations

`Switch#delete flash:vlan.dat`	Removes entire VLAN database from flash
	WARNING: Make sure there is *no* space between the colon (:) and the characters *vlan.dat*. You can potentially erase the entire contents of the flash with this command if the syntax is not correct. Make sure you read the output from the switch. If you need to cancel, press Ctrl-c to escape back to privileged mode: (Switch#) Switch#**delete flash:vlan.dat** Delete filename [vlan.dat]? Delete flash:vlan.dat? [confirm] Switch#
`Switch(config)#interface fastethernet 0/5`	Moves to interface config mode
`Switch(config-if)#no switchport access vlan 5`	Removes port from VLAN 5 and reassigns it to VLAN 1—the default VLAN
`Switch(config-if)#exit`	Moves to global config mode
`Switch(config)#no vlan 5`	Removes VLAN 5 from the VLAN database
or	
`Switch#vlan database`	Enters VLAN Database mode
`Switch(vlan)#no vlan 5`	Removes VLAN 5 from the VLAN database
`Switch(vlan)#exit`	Applies changes, increases the revision number by 1, and exits VLAN Database mode

NOTE: When you delete a VLAN from a switch that is in VTP server mode, the VLAN is removed from the VLAN database for all switches in the VTP domain. When you delete a VLAN from a switch that is in VTP transparent mode, the VLAN is deleted only on that specific switch.

NOTE: You cannot delete the default VLANs for the different media types: Ethernet VLAN 1 and FDDI or Token Ring VLANs 1002 to 1005.

CAUTION: When you delete a VLAN, any ports assigned to that VLAN become inactive. They remain associated with the VLAN (and thus inactive) until you assign them to a new VLAN. Therefore, it is recommended that you reassign ports to a new VLAN or the default VLAN before you delete a VLAN from the VLAN database.

Verifying VLAN Trunking

`Switch#show interface fastethernet 0/1 switchport`	Displays the administrative and operational status of a trunking port

VLAN Trunking Protocol

VLAN Trunking Protocol (VTP) is a Cisco proprietary protocol that allows for VLAN configuration (addition, deletion, or renaming of VLANS) to be consistently maintained across a common administrative domain.

Using Global Configuration Mode

`Switch(config)#vtp mode client`	Changes the switch to VTP client mode
`Switch(config)#vtp mode server`	Changes the switch to VTP server mode
`Switch(config)#vtp mode transparent`	Changes the switch to VTP transparent mode
	NOTE: By default, all Catalyst switches are in server mode.
`Switch(config)#no vtp mode`	Returns the switch to the default VTP server mode
`Switch(config)#vtp domain domain-name`	Configures the VTP domain name. The name can be from 1 to 32 characters long.

	NOTE: All switches operating in VTP server or client mode must have the same domain name to ensure communication.
`Switch(config)#`**`vtp password`** `password`	Configures a VTP password. In Cisco IOS Software Release 12.3 and later, the password is an ASCII string from 1 to 32 characters long. If you are using a Cisco IOS release earlier than 12.3, the password length ranges from 8 to 64 characters long.
	NOTE: To communicate with each other, all switches must have the same VTP password set.
`Switch(config)#`**`vtp v2-mode`**	Sets the VTP domain to Version 2. This command is for Cisco IOS Software Release 12.3 and later. If you are using a Cisco IOS release earlier than 12.3, the command is **vtp version 2**.
	NOTE: VTP Versions 1 and 2 are not interoperable. All switches must use the same version. The biggest difference between Versions 1 and 2 is that Version 2 has support for Token Ring VLANs.
`Switch(config)#`**`vtp pruning`**	Enables VTP pruning
	NOTE: By default, VTP pruning is disabled. You need to enable VTP pruning on only one switch in VTP server mode.

NOTE: Only VLANs included in the pruning-eligible list can be pruned. VLANs 2 through 1001 are pruning eligible by default on trunk ports. Reserved VLANs and extended-range VLANs cannot be pruned. To change which eligible VLANs can be pruned, use the interface-specific **switchport trunk pruning vlan** command:

```
Switch(config-if)#switchport trunk pruning vlan remove 4, 20-30
! Removes VLANs 4 and 20-30
Switch(config-if)#switchport trunk pruning vlan except 40-50
! All VLANs are added to the pruning list except for 40-50
```

Using VLAN Database Mode

The VLAN Database mode has been deprecated and will be removed in some future Cisco IOS release. Recommended practice dictates using only the VLAN-configuration mode.

Do not use

Switch#**vlan database**	Enters VLAN Database mode
Switch(vlan)#**vtp client**	Changes the switch to VTP client mode
Switch(vlan)#**vtp server**	Changes the switch to VTP server mode
Switch(vlan)#**vtp transparent**	Changes the switch to VTP transparent mode
	NOTE: By default, all Catalyst switches are in server mode.
Switch(vlan)#**vtp domain** *domain-name*	Configures the VTP domain name. The name can be from 1 to 32 characters long.
	NOTE: All switches operating in VTP server or client mode must have the same domain name to ensure communication.
Switch(vlan)#**vtp password** *password*	Configures a VTP password. In Cisco IOS Release 12.3 and later, the password is an ASCII string from 1 to 32 characters long. If you are using a Cisco IOS release earlier than IOS 12.3, the password length ranges from 8 to 64 characters long.
	NOTE: All switches must have the same VTP password set in order to communicate with each other
Switch(vlan)#**vtp v2-mode**	Sets the VTP domain to Version 2. This command is for Cisco IOS Release 12.3 and later. If you are using a Cisco IOS release earlier than 12.3, the command is **vtp version 2**.
	NOTE: VTP Versions 1 and 2 are not interoperable. All switches must use the same version. The biggest difference between Versions 1 and 2 is that Version 2 has support for Token Ring VLANs.
Switch(vlan)#**vtp pruning**	Enables VTP pruning.

	NOTE: By default, VTP pruning is disabled. You need to enable VTP pruning on only one switch in VTP server mode.
	NOTE: Only VLANs included in the pruning-eligible list can be pruned. VLANs 2 through 1001 are pruning eligible by default on trunk ports. Reserved VLANs and extended-range VLANs cannot be pruned. To change which eligible VLANs can be pruned, use the interface-specific **switchport trunk pruning vlan** command: Switch(config-if)#**switchport trunk pruning vlan remove 4, 20-30** ! Removes VLANs 4 and 20-30 Switch(config-if)#**switchport trunk pruning vlan except 40-50** All VLANs are added to the pruning list except for 40 through 50.
Switch(vlan)#**exit**	Applies changes to VLAN database, increases the revision number by 1, and exits back to privileged mode

Verifying VTP

Switch#**show vtp status**	Displays general information about VTP configuration
Switch#**show vtp counters**	Displays the VTP counters for the switch

NOTE: If trunking has been established before VTP is set up, VTP information is propagated throughout the switch fabric almost immediately. However, because VTP information is advertised only every 300 seconds (5 minutes) unless a change has been made to force an update, it can take several minutes for VTP information to be propagated.

Configuration Example: VLANs

Figure 2-1 shows the network topology for the configuration that follows, which shows how to configure VLANs using the commands covered in this chapter.

Figure 2-1   Network Topology for VLAN Configuration Example

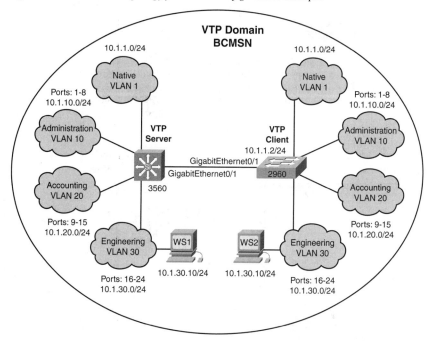

3560 Switch

Switch>`enable`	Moves to privileged mode
Switch#`configure terminal`	Moves to global configuration mode
Switch(config)#`hostname 3560`	Sets the host name
3560(config)#`vtp mode server`	Changes the switch to VTP server mode. Note that server is the default setting for a 3560 switch.
3560(config)#`vtp domain bcmsn`	Configures the VTP domain name to bcmsn
3560(config)#`vtp password tower`	Sets the VTP password to tower

`3560(config)#vlan 10`	Creates VLAN 10 and enters VLAN-configuration mode
`3560(config-vlan)#name Admin`	Assigns a name to the VLAN
`3560(config-vlan)#exit`	Increases the revision number by 1 and returns to global configuration mode
`3560(config)#vlan 20`	Creates VLAN 20 and enters VLAN-configuration mode
`3560(config-vlan)#name Accounting`	Assigns a name to the VLAN
`3560(config-vlan)#vlan 30`	Creates VLAN 30 and enters VLAN-configuration mode. Note that you do not have to exit back to global configuration mode to execute this command.
`3560(config-vlan)#name Engineering`	Assigns a name to the VLAN
`3560(config-vlan)#exit`	Increases the revision number by 1 and returns to global configuration mode
`3560(config)#interface range fasthethernet 0/1 - 8`	Enables you to set the same configuration parameters on multiple ports at the same time
`3560(config-if-range)#switchport mode access`	Sets ports 1–8 as access ports
`3560(config-if-range)#switchport access vlan 10`	Assigns ports 1–8 to VLAN 10
`3560(config-if-range)#interface range fastethernet 0/9 - 15`	Enables you to set the same configuration parameters on multiple ports at the same time
`3560(config-if-range)#switchport mode access`	Sets ports 9–15 as access ports
`3560(config-if-range)#switchport access vlan 20`	Assigns ports 9–15 to VLAN 20
`3560(config-if-range)#interface range fastethernet 0/16 - 24`	Enables you to set the same configuration parameters on multiple ports at the same time
`3560(config-if-range)#switchport mode access`	Sets ports 16–24 as access ports

`3560(config-if-range)#`**`switchport access vlan 30`**	Assigns ports 16–24 to VLAN 30
`3560(config-if-range)#`**`exit`**	Returns to global configuration mode
`3560(config)#`**`interface gigabitethernet 0/1`**	Moves to interface configuration mode
`3560(config-if)#`**`switchport trunk encapsulation dot1q`**	Specifies 802.1Q encapsulation on the trunk link
`3560(config-if)#`**`switchport mode trunk`**	Puts the interface into permanent trunking mode and negotiates to convert the link into a trunk link
`3560(config-if)#`**`exit`**	Returns to global configuration mode
`3560(config)#`**`exit`**	Returns to privileged mode
`3560#`**`copy running-config startup-config`**	Saves the configuration in NVRAM

2960 Switch

`Switch>`**`enable`**	Moves to privileged mode
`Switch#`**`configure terminal`**	Moves to global configuration mode
`Switch(config)#`**`hostname 2960`**	Sets the host name
`2960(config)#`**`vtp mode client`**	Changes the switch to VTP client mode
`2960(config)#`**`vtp domain bcmsn`**	Configures the VTP domain name to bcmsn
`2960(config)#`**`interface range fastethernet 0/1 - 8`**	Enables you to set the same configuration parameters on multiple ports at the same time
`2960(config-if-range)#`**`switchport mode access`**	Sets ports 1–8 as access ports
`2960(config-if-range)#`**`switchport access vlan 10`**	Assigns ports 1–8 to VLAN 10

2960(config-if-range)#**interface range fastethernet 0/9 - 15**	Enables you to set the same configuration parameters on multiple ports at the same time
2960(config-if-range)#**switchport mode access**	Sets ports 9–15 as access ports
2960(config-if-range)#**switchport access vlan 20**	Assigns ports 9–15 to VLAN 20
2960(config-if-range)#**interface range fastethernet 0/16 - 24**	Enables you to set the same configuration parameters on multiple ports at the same time
2960(config-if-range)#**switchport mode access**	Sets ports 16–24 as access ports
2960(config-if-range)#**switchport access vlan 30**	Assigns ports 16–24 to VLAN 30
2960(config-if-range)#**exit**	Returns to global configuration mode
2960(config)#**int gigabitethernet 0/1**	Moves to interface configuration mode
2960(config-if)#**switchport mode trunk**	Puts the interface into permanent trunking mode and negotiates to convert the link into a trunk link
2960(config-if)#**exit**	Returns to global configuration mode
2960(config)#**exit**	Returns to privileged mode
2960#**copy running-config startup-config**	Saves the configuration in NVRAM

This chapter provides information and commands concerning the following topics:

Spanning Tree Protocol

- Enabling Spanning Tree Protocol (STP)
- Configuring the root switch
- Configuring a secondary root switch
- Configuring port priority
- Configuring the path cost
- Configuring the switch priority of a VLAN
- Configuring STP timers
- Verifying STP
- Optional STP configurations
  — PortFast
  — BPDU Guard
  — BPDU Filtering
  — UplinkFast
  — BackboneFast
  — Root Guard
  — Loop Guard
  — Unidirectional Link Detection (UDLD)
- Changing the spanning-tree mode
- Extended System ID
- Enabling Rapid Spanning Tree
- Enabling Multiple Spanning Tree
- Verifying MST
- Troubleshooting STP

EtherChannel

- Interface modes in EtherChannel
  — Without Port aggregation protocol (PAgP) or Link Aggregation Control Protocol (LACP)
  — With PagP
  — With LACP
- Guidelines for configuring EtherChannel
- Configuring L2 EtherChannel

- Configuring L3 EtherChannel
- Verifying EtherChannel
- Configuring EtherChannel load balancing
- Types of EtherChannel load balancing
- Verifying EtherChannel load balancing

Spanning Tree Protocol

Enabling Spanning Tree Protocol

Switch(config)#**spanning-tree vlan 5**	Enables STP on VLAN 5
Switch(config)#**no spanning-tree vlan 5**	Disables STP on VLAN 5

NOTE: If more VLANs are defined in the VLAN Trunking Protocol (VTP) than there are spanning-tree instances, you can only have STP on 64 VLANs. If you have more than 128 VLANs, it is recommended that you use Multiple STP.

Configuring the Root Switch

Switch(config)#**spanning-tree vlan 5 root**	Modifies the switch priority from the default 32768 to a lower value to allow the switch to become the root switch for VLAN 5
	NOTE: If all other switches have extended system ID support, this switch resets its priority to 24576. If any other switch has a priority set to below 24576 already, this switch sets its own priority to 4096 *less* than the lowest switch priority. If by doing this the switch would have a priority of less than 1, this command fails.
Switch(config)#**spanning-tree vlan 5 root primary**	Switch recalculates timers along with priority to allow the switch to become the root switch for VLAN 5
	TIP: The root switch should be a backbone or distribution switch.
Switch(config)#**spanning-tree vlan 5 root primary diameter 7**	Configures the switch to be the root switch for VLAN 5 and sets the network diameter to 7

	TIP: The **diameter** keyword is used to define the maximum number of switches between any two end stations. The range is from 2 to 7 switches.
`Switch(config)#spanning-tree vlan 5 root primary hello-time 4`	Configures the switch to be the root switch for VLAN 5 and sets the hello-delay timer to 4 seconds
	TIP: The **hello-time** keyword sets the hello-delay timer to any amount between 1 and 10 seconds. The default time is 2 seconds.

Configuring a Secondary Root Switch

`Switch(config)#spanning-tree vlan 5 root secondary`	Switch recalculates timers along with priority to allow the switch to become the root switch for VLAN 5 should the primary root switch fail
	NOTE: If all other switches have extended system ID support, this switch resets its priority to 28672. Therefore, if the root switch fails, and all other switches are set to the default priority of 32768, this becomes the new root switch. For switches without Extended System ID support, the switch priority is changed to 16384.
`Switch(config)#spanning-tree vlan 5 root secondary diameter 7`	Configures the switch to be the secondary root switch for VLAN 5 and sets the network diameter to 7
`Switch(config)#spanning-tree vlan 5 root secondary hello-time 4`	Configures the switch to be the secondary root switch for VLAN 5 and sets the hello-delay timer to 4 seconds

Configuring Port Priority

`Switch(config)#interface gigabitethernet 0/1`	Moves to interface configuration mode
`Switch(config-if)#spanning-tree port-priority 64`	Configures the port priority for the interface that is an access port

`Switch(config-if)#spanning-tree vlan 5 port-priority 64`	Configures the VLAN port priority for an interface that is a trunk port
	NOTE: Port priority is used to break a tie when two switches have equal priorities for determining the root switch. The number can be between 0 and 255. The default port priority is 128. The lower the number, the higher the priority.

Configuring the Path Cost

`Switch(config)#interface gigabitethernet 0/1`	Moves to interface config mode
`Switch(config-if)#spanning-tree cost 100000`	Configures the cost for the interface that is an access port
`Switch(config-if)#spanning-tree vlan 5 cost 1000000`	Configures the VLAN cost for an interface that is a trunk port
	NOTE: If a loop occurs, STP uses the path cost when trying to determine which interface to place into the forwarding state. A higher path cost means a lower speed transmission. The range of the **cost** keyword is 1 through 200000000. The default is based on the media speed of the interface.

Configuring the Switch Priority of a VLAN

`Switch(config)#spanning-tree vlan 5 priority 12288`	Configures the switch priority of VLAN 5 to 12288

NOTE: With the **priority** keyword, the range is 0 to 61440 in increments of 4096. The default is 32768. The lower the priority, the more likely the switch will be chosen as the root switch.

Only the following numbers can be used as a priority value:

0	4096	8192	12288
16384	20480	24576	28672
32768	36864	40960	45056
49152	53248	57344	61440

CAUTION: Cisco recommends caution when using this command. Cisco further recommends that the **spanning-tree vlan** *x* **root primary** or the **spanning-tree vlan** *x* **root secondary** command be used instead to modify the switch priority.

Configuring STP Timers

`Switch(config)#spanning-tree vlan 5 hello-time 4`	Changes the hello-delay timer to 4 seconds on VLAN 5
`Switch(config)#spanning-tree vlan 5 forward-time 20`	Changes the forward-delay timer to 20 seconds on VLAN 5
`Switch(config)#spanning-tree vlan 5 max-age 25`	Changes the maximum-aging timer to 25 seconds on VLAN 5

NOTE: For the **hello-time** command, the range is 1 to 10 seconds. The default is 2 seconds.

For the **forward-time** command, the range is 4 to 30 seconds. The default is 15 seconds.

For the **max-age** command, the range is 6 to 40 seconds. The default is 20 seconds.

CAUTION: Cisco recommends caution when using this command. Cisco further recommends that the **spanning-tree vlan** *x* **root primary** or the **spanning-tree vlan** *x* **root secondary** command be used instead to modify the switch timers.

Verifying STP

`Switch#show spanning-tree`	Displays STP information
`Switch#show spanning-tree active`	Displays STP information on active interfaces only
`Switch#show spanning-tree brief`	Displays a brief status of the STP
`Switch#show spanning-tree detail`	Displays a detailed summary of interface information
`Switch#show spanning-tree interface gigabitethernet 0/1`	Displays STP information for interface gigabitethernet 0/1
`Switch#show spanning-tree summary`	Displays a summary of port states
`Switch#show spanning-tree summary totals`	Displays the total lines of the STP section
`Switch#show spanning-tree vlan 5`	Displays STP information for VLAN 5

Optional STP Configurations

Although the following commands are not mandatory for STP to work, you might find these helpful in fine-tuning your network.

PortFast

Switch(config)#**interface fastethernet 0/10**	Moves to interface config mode
Switch(config-if)#**spanning-tree portfast**	Enables PortFast on an access port
Switch(config-if)#**spanning-tree portfast trunk**	Enables PortFast on a trunk port
	WARNING: Use the **portfast** command only when connecting a single end station to an access or trunk port. Using this command on a port connected to a switch or hub could prevent spanning tree from detecting loops.
	NOTE: If you enable the voice VLAN feature, PortFast is enabled automatically. If you disable voice VLAN, PortFast is still enabled.
Switch#**show spanning-tree interface fastethernet 0/10 portfast**	Displays PortFast information on interface fastethernet 0/10

(handwritten: WHY? / SHOW RUN-CONF INT F0/7 (VERIFIES PORTFAST))

BPDU Guard

Switch(config)#**spanning-tree portfast bpduguard default**	Globally enables BPDU Guard
Switch(config)#**interface range fastethernet 0/1 - 5**	Enters interface range configuration mode
Switch(config-if-range)#**spanning-tree portfast**	Enables PortFast on all interfaces in the range
	NOTE: By default, BPDU Guard is disabled.
Switch(config)#**errdisable recovery cause bpduguard**	Allows port to reenable itself if the cause of the error is BPDU Guard by setting a recovery timer

`Switch(config)#`**`errdisable`** **`recovery interval 400`**	Sets recovery timer to 400 seconds. Default is 300 seconds. Range is from 30 to 86400 seconds
`Switch#`**`show spanning-tree`** **`summary totals`**	Verifies whether BPDU Guard is enabled or disabled
`Switch#`**`show errdisable recovery`**	Displays errdisable recovery timer information

BPDU Filtering

`Switch(config)#`**`spanning-tree`** **`portfast bpdufilter default`**	Globally enables BPDU Filtering— prevents ports in PortFast from sending or receiving bridge protocol data units (BPDU)
`Switch(config)#`**`interface range`** **`fastethernet 0/1 - 5`**	Enters interface range config mode
`Switch(config-if-range)#`**`spanning-`** **`tree portfast`**	Enables PortFast on all interfaces in the range
	NOTE: By default, BPDU Filtering is disabled.
	CAUTION: Enabling BPDU Filtering on an interface, or globally, is the same as disabling STP, which can result in spanning-tree loops being created but not detected.
`Switch(config)#`**`interface`** **`fastethernet 0/15`**	Moves to interface config mode
`Switch(config-if)#`**`spanning-tree`** **`bpdufilter enable`**	Enables BPDU Filtering on the interface without enabling the PortFast feature
`Switch#`**`show running-config`**	Verifies BPDU Filtering is enabled on interfaces

UplinkFast

Switch(config)#**spanning-tree uplinkfast**	Enables UplinkFast
Switch(config)#**spanning-tree uplinkfast max-update-rate 200**	Enables UplinkFast and sets the update packet rate to 200 packets/second
	NOTE: UplinkFast cannot be set on an individual VLAN. The **spanning-tree uplinkfast** command affects all VLANs.
	NOTE: For the **max-update-rate** argument, the range is 0 to 32000 packets/second. The default is 150. If you set the rate to 0, station-learning frames are not generated. This will cause STP to converge more slowly after a loss of connectivity.
Switch#**show spanning-tree summary**	Verifies UplinkFast has been enabled

NOTE: UplinkFast cannot be enabled on VLANs that have been configured for switch priority.

NOTE: UplinkFast is most useful in access-layer switches, or switches at the edge of the network. It is not appropriate for backbone devices.

BackboneFast

Switch(config)#**spanning-tree backbonefast**	Enables BackboneFast
Switch#**show spanning-tree summary**	Verifies BackboneFast has been enabled

Root Guard

Switch(config)#**interface fastethernet 0/1**	Moves to interface config mode
Switch(config-if)#**spanning-tree guard root**	Enables Root Guard on the interface
Switch#**show running-config**	Verifies Root Guard is enabled on the interface

NOTE: You cannot enable both Root Guard and Loop Guard at the same time.

NOTE: Root Guard enabled on an interface applies to all VLANs to which the interface belongs.

NOTE: Do not enable Root Guard on interfaces to be used by the UplinkFast feature.

Loop Guard

Switch#**show spanning-tree active**	Shows which ports are alternate or root ports
Switch#**show spanning-tree mst**	Shows which ports are alternate or root ports
Switch#**configure terminal**	Moves to global configuration mode
Switch(config)#**spanning-tree loopguard default**	Enables Loop Guard globally on the switch
Switch(config)#**exit**	Returns to privileged mode
Switch#**show running-config**	Verifies that Loop Guard has been enabled

NOTE: You cannot enable both Root Guard and Loop Guard at the same time.

NOTE: This feature is most effective when it is configured on the entire switched network.

NOTE: Loop Guard operates only on ports that are considered to be point to point by the STP.

Unidirectional Link Detection

Switch(config)#**udld enable**	Enables Unidirectional Link Detection (UDLD) on all fiber-optic Interfaces
	NOTE: By default, UDLD is disabled.
Switch(config)#**interface fastethernet 0/24**	Moves to interface config mode

Switch(config-if)#`udld port`	Enables UDLD on this interface—required for copper-based interfaces
	NOTE: On a fiber-optic (FO) interface, the interface command **udld port** overrides the global command **udld enable**. Therefore, if you issue the command **no udld port** on an FO interface, you will still have the globally enabled **udld enable** command to deal with.
Switch#`show udld`	Displays UDLD information
Switch#`show udld interface fastethernet 0/1`	Displays UDLD information for interface fastethernet 0/1
Switch#`udld reset`	Resets all interfaces shut down by UDLD
	NOTE: You can also use the **shutdown** command, followed by a **no shutdown** command in interface config mode, to restart a disabled interface.

Changing the Spanning-Tree Mode

There are different types of spanning tree that can be configured on a Cisco switch. The options vary according to the platform:

- **Per-VLAN Spanning Tree (PVST)**—There is one instance of spanning tree for each VLAN. This is a Cisco proprietary protocol.
- **Per-VLAN Spanning Tree Plus (PVST+)**—Also Cisco proprietary. Has added extensions to the PVST protocol.
- **Rapid PVST+**—This mode is the same as PVST+ except that it uses a rapid convergence based on the 802.1w standard.
- **Multiple Spanning Tree Protocol (MSTP)**—IEEE 802.1s. Extends the 802.1w Rapid Spanning Tree (RST) algorithm to multiple spanning trees. Multiple VLANs can map to a single instance of RST. You cannot run MSTP and PVST at the same time.

Switch(config)#`spanning-tree mode mst`	Enables MSTP. This command is available only on a switch running the EI software image.
Switch(config)#`spanning-tree mode pvst`	Enables PVST—this is the default setting
Switch(config)#`spanning-tree mode rapid-pvst`	Enables Rapid PVST+

Extended System ID

`Switch(config)#`**`spanning-tree extend`** **`system-id`**	Enables Extended System ID, also known as MAC Address Reduction
	NOTE: Catalyst switches running software earlier than Cisco IOS Release 12.1(8)EA1 do not support the Extended System ID.
`Switch#`**`show spanning-tree summary`**	Verifies Extended System ID is enabled
`Switch#`**`show running-config`**	Verifies Extended System ID is enabled

Enabling Rapid Spanning Tree

`Switch(config)#`**`spanning-tree mode`** **`rapid-pvst`**	Enables Rapid PVST+
`Switch(config)#`**`interface`** **`fastethernet 0/1`**	Moves to interface config mode
`Switch(config-if)#`**`spanning-tree`** **`link-type point-to-point`**	Sets the interface to be a point-to-point interface
	NOTE: By setting the link type to point-to-point, this means that if you connect this port to a remote port, and this port becomes a designated port, the switch will negotiate with the remote port and transition the local port to a forwarding state.
`Switch(config-if)#`**`exit`**	
`Switch(config)#`**`clear spanning-tree`** **`detected-protocols`**	
	NOTE: The **clear spanning-tree detected-protocols** command restarts the protocol migration process on the switch if any port is connected to a port on a legacy 802.1D switch.

Enabling Multiple Spanning Tree

`Switch(config)#`**`spanning-tree mst`** **`configuration`**	Enters MST config mode
`Switch(config-mst)#`**`instance 1 vlan 4`**	Maps VLAN 4 to an Multiple Spanning Tree (MST) instance
`Switch(config-mst)#`**`instance 1 vlan`** **`1-15`**	Maps VLANs 1–15 to MST instance 1
`Switch(config-mst)#`**`instance 1 vlan`** **`10,20,30`**	Maps VLANs 10, 20, and 30 to MST instance 1
	NOTE: For the **instance** *x* **vlan** *y* command, the instance must be a number between 1 and 15, and the VLAN range is 1 to 4094.
`Switch(config-mst)#`**`name region12`**	Specifies the configuration name to be region12.
	NOTE: The **name** argument can be up to 32 characters long and is case sensitive.
`Switch(config-mst)#`**`revision 4`**	Specifies the revision number
	NOTE: The range for the **revision** argument is 0 to 65535.
`Switch(config-mst)#`**`show pending`**	Verifies the configuration by displaying a summary of what you have configured for the MST region
`Switch(config-mst)#`**`exit`**	Applies all changes and returns to global config mode
`Switch(config)#`**`spanning-tree mst 1`**	Enables MST
	CAUTION: Changing spanning-tree modes can disrupt traffic because all spanning-tree instances are stopped for the old mode and restarted in the new mode.
	NOTE: You cannot run both MSTP and PVST at the same time.
`Switch(config)#`**`spanning-tree mst 1`** **`root primary`**	Configures a switch as a primary root switch within MST instance 1. The primary root switch priority is 24576.

Switch(config)#**spanning-tree mst 1 root secondary**	Configures a switch as a secondary root switch within MST instance 1. The secondary root switch priority is 28672.
Switch(config)#**exit**	Returns to privileged mode

Verifying MST

Switch#**show spanning-tree mst configuration**	Displays the MST region configuration
Switch#**show spanning-tree mst 1**	Displays the MST information for instance 1
Switch#**show spanning-tree mst interface fastethernet 0/1**	Displays the MST information for interface fastethernet 0/1
Switch#**show spanning-tree mst 1 interface fastethernet 0/1**	Displays the MST information for instance 1 on interface fastethernet 0/1
Switch#**show spanning-tree mst 1 detail**	Shows detailed information about MST instance 1

Troubleshooting Spanning Tree

Switch#**debug spanning-tree all**	Displays all spanning-tree debugging events
Switch#**debug spanning-tree events**	Displays spanning-tree debugging topology events
Switch#**debug spanning-tree backbonefast**	Displays spanning-tree debugging BackboneFast events
Switch#**debug spanning-tree uplinkfast**	Displays spanning-tree debugging UplinkFast event
Switch#**debug spanning-tree mstp all**	Displays all MST debugging events
Switch#**debug spanning-tree switch state**	Displays spanning-tree port state changes
Switch#**debug spanning-tree pvst+**	Displays PVST+ events

SHOW SWITCHPORT CAPABILITIES. —CHECKS FOR ETHERCHANNEL
FEATURE.

Configuration Example: STP

Figure 3-1 shows the network topology for the configuration that follows, which shows how to configure STP using commands covered in this chapter.

Figure 3-1   Network Topology for STP Configuration Example

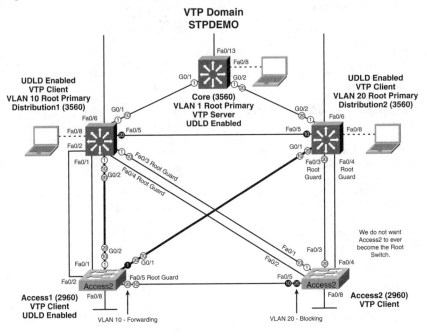

Core Switch (3560)

`Switch>`**`enable`**	Moves to privileged mode
`Switch#`**`configure terminal`**	Moves to global config mode
`Switch(config)#`**`hostname Core`**	Sets host name
`Core(config)#`**`no ip domain-lookup`**	Turns off Dynamic Name System (DNS) queries so that spelling mistakes will not slow you down
`Core(config)#`**`vtp mode server`**	Changes the switch to VTP server mode. This is the default mode.

`Core(config)#vtp domain stpdemo`	Configures the VTP domain name to stpdemo
`Core(config)#vlan 10`	Creates VLAN 10 and enters VLAN-config mode
`Core(config-vlan)#name Accounting`	Assigns a name to the VLAN
`Core(config-vlan)#exit`	Returns to global config mode
`Core(config)#vlan 20`	Creates VLAN 20 and enters VLAN-config mode
`Core(config-vlan)#name Marketing`	Assigns a name to the VLAN
`Core(config-vlan)#exit`	Returns to global config mode
`Core(config)#spanning-tree vlan 1 root primary`	Configures the sitch to become the root switch for VLAN 1
`Core(config)#udld enable`	Enables UDLD
`Core(config)#exit`	Returns to privileged mode
`Core#copy running-config startup-config`	Saves the configuration to NVRAM

Distribution 1 Switch (3560)

`Switch>enable`	Moves to privileged mode
`Switch#configure terminal`	Moves to global config mode
`Switch(config)#hostname Distribution1`	Sets host name
`Distribution1(config)#no ip domain-lookup`	Turns off DNS queries so that spelling mistakes will not slow you down
`Distribution1(config)#vtp domain stpdemo`	Configures the VTP domain name to stpdemo
`Distribution1(config)#vtp mode client`	Changes the switch to VTP client mode
`Distribution1(config)#spanning-tree vlan 10 root primary`	Configures the switch to become the root switch of VLAN 10
`Distribution1(config)#udld enable`	Enables UDLD on all FO interfaces

`Distribution1(config)#`**`interface range fastethernet 0/3 - 4`**	Moves to interface range mode
`Distribution1(config-if)#`**`spanning-tree guard root`**	Prevents switch on the other end of the link (Access2) from becoming the root switch
`Distribution1(config-if)#`**`exit`**	Returns to global config mode
`Distribution1(config)#`**`exit`**	Returns to privileged mode
`Distribution1#`**`copy running-config startup-config`**	Saves the configuration to NVRAM

Distribution 2 Switch (3560)

`Switch>`**`enable`**	Moves to privileged mode
`Switch#`**`configure terminal`**	Moves to global config mode
`Switch(config)#`**`hostname Distribution2`**	Sets host name
`Distribution2(config)#`**`no ip domain-lookup`**	Turns off DNS queries so that spelling mistakes will not slow you down
`Distribution2(config)#`**`vtp domain stpdemo`**	Configures the VTP domain name to stpdemo
`Distribution2(config)#`**`vtp mode client`**	Changes the switch to VTP client mode
`Distribution2(config)#`**`spanning-tree vlan 20 root primary`**	Configures the switch to become the root switch of VLAN 20
`Distribution2(config)#`**`udld enable`**	Enables UDLD on all FO interfaces
`Distribution2(config)#`**`interface range fastethernet 0/3 - 4`**	Moves to interface range mode
`Distribution2(config-if)#`**`spanning-tree guard root`**	Prevents the switch on the other end of link (Access2) from becoming the root switch
`Distribution2(config-if)#`**`exit`**	Returns to global config mode
`Distribution2(config)#`**`exit`**	Returns to privileged mode
`Distribution2#`**`copy running-config startup-config`**	Saves the configuration to NVRAM

Access 1 Switch (2960)

`Switch>`**`enable`**	Moves to privileged mode
`Switch#`**`configure terminal`**	Moves to global config mode
`Switch(config)#`**`hostname Access1`**	Sets host name
`Access1(config)#`**`no ip domain-lookup`**	Turns off DNS queries so that spelling mistakes will not slow you down
`Access1(config)#`**`vtp domain stpdemo`**	Configures the VTP domain name to *stpdemo*
`Access1(config)#`**`vtp mode client`**	Changes the switch to VTP client mode
`Access1(config)#`**`interface range fastethernet 0/6 - 12`**	Moves to interface range config mode
`Access1(config-if-range)#`**`switchport mode access`**	Places all interfaces in access mode
`Access1(config-if-range)#`**`spanning-tree portfast`**	Places all ports directly into forwarding mode
`Access1(config-if-range)#`**`spanning-tree bpduguard enable`**	Enables BPDU Guard
`Access1(config-if-range)#`**`exit`**	Moves back to global config mode
`Access1(config)#`**`spanning-tree uplinkfast`**	Enables UplinkFast to reduce STP convergence time
`Access1(config)#`**`interface fastethernet 0/5`**	Moves to interface config mode
`Access1(config-if)#`**`spanning-tree guard root`**	Prevents the switch on the other end of link (Access2) from becoming the root switch
`Access1(config-if)#`**`exit`**	Returns to global config mode
`Access1(config)#`**`udld enable`**	Enables UDLD on all FO interfaces
`Access1(config)#`**`exit`**	Returns to privileged mode
`Access1#`**`copy running-config startup-config`**	Saves the configuration to NVRAM

Access 2 Switch (2960)

`Switch>`**`enable`**	Moves to privileged mode
`Switch#`**`configure terminal`**	Moves to global config mode
`Switch(config)#`**`hostname Access2`**	Sets host name
`Access2(config)#`**`no ip domain-lookup`**	Turns off DNS queries so that spelling mistakes will not slow you down
`Access2(config)#`**`vtp domain stpdemo`**	Configures the VTP domain name to stpdemo
`Access2(config)#`**`vtp mode client`**	Changes the switch to VTP client mode
`Access2(config)#`**`interface range fastethernet 0/6 - 12`**	Moves to interface range config mode
`Access2(config-if-range)#`**`switchport mode access`**	Places all interfaces in access mode
`Access2(config-if-range)#`**`spanning-tree portfast`**	Places all ports directly into forwarding mode
`Access2(config-if-range)#`**`spanning-tree bpduguard enable`**	Enables BPDU Guard
`Access2(config-if-range)#`**`exit`**	Moves back to global config mode
`Access2(config)#`**`spanning-tree vlan 10 priority 61440`**	Ensures this switch will not become the root switch for VLAN 10
`Access2(config)#`**`exit`**	Returns to privileged mode
`Access2#`**`copy running-config startup-config`**	Saves config to NVRAM

EtherChannel

EtherChannel provides fault-tolerant high-speed links between switches, routers, and servers. An EtherChannel consists of individual Fast Ethernet or Gigabit Ethernet links bundled into a single logical link. If a link within an EtherChannel fails, traffic previously carried over that failed link changes to the remaining links within the EtherChannel.

- CREATES HIGH B/W LOGICAL LINKS
- LOAD BALANCES AMONG PHYSICAL LINKS INVOLVED.
- PROVIDES AUTO FAILOVER
- SIMPLIFIES SUBSEQUENT LOGICAL CONFIGURATIONS

Interface Modes in EtherChannel

Mode	Protocol	Description
On	None	Forces the interface into an EtherChannel without PAgP or LACP. Channel only exists if connected to another interface group also in On mode.
Auto	PAgP	Places the interface into a passive negotiating state—will respond to PAgP packets, but will not initiate PAgP negotiation
Desirable	PAgP	Places the interface into an active negotiating state—will send PAgP packets to start negotiations
Passive	LACP	Places the interface into a passive negotiating state—will respond to LACP packets, but will not initiate LACP negotiation
Active	LACP	Place the interface into an active negotiating state—will send LACP packets to start negotiations

Guidelines for Configuring EtherChannel

- PAgP is Cisco proprietary.
- LACP is defined in 802.3ad.
- Can combine from two to eight parallel links.
- All ports must be identical:
  — Same speed and duplex
  — Cannot mix Fast Ethernet and Gigabit Ethernet
  — Cannot mix PAgP and LACP
  — Must all be VLAN trunk or nontrunk operational status
- All links must be either L2 or L3 in a single channel group.
- To create a channel in PAgP, sides must be set to
  — Auto-Desirable
  — Desirable-Desirable
- To create a channel in LACP, sides must be set to
  — Active-Active
  — Active-Passive
- To create a channel without using PAgP or LACP, sides must be set to On-On.
- Do *not* configure a GigaStack Gigabit Interface Converter (GBIC) as part of an EtherChannel.
- An interface that is already configured to be a Switched Port Analyzer (SPAN) destination port will not join an EtherChannel group until SPAN is disabled.
- Do *not* configure a secure port as part of an EtherChannel.

- Interfaces with different native VLANs cannot form an EtherChannel.
- When using trunk links, ensure all trunks are in the same mode—Inter-Switch Link (ISL) or Dot1Q.

Configuring L2 EtherChannel

`Switch(config)#interface range fastethernet 0/1 - 4`	Moves to interface range config mode
`Switch(config-if-range)#channel-protocol pagp`	Specifies the PAgP protocol to be used in this channel
`or`	
`Switch(config-if-range)#channel-protocol lacp`	Specifies the LACP protocol to be used in this channel
`Switch(config-if-range)#channel-group 1 mode {desirable ¦ auto ¦ on ¦ passive ¦ active }`	Creates channel group 1 and assigns interfaces 01–04 as part of it. Use whichever mode is necessary, depending on your choice of protocol.

Configuring L3 EtherChannel

`3560Switch(config)#interface port-channel 1`	Creates the port-channel logical *(LAYER 2)* interface, and moves to interface config mode. Valid channel numbers are 1–48.
`3560Switch(config-if)#no switchport`	Puts the interface into Layer 3 mode *I MPORTANT. ENABLES USE OF IP.*
`3560Switch(config-if)#ip address 172.16.10.1 255.255.255.0`	Assigns IP address and netmask
`3560Switch(config-if)#exit`	Moves to global config mode
`3560Switch(config)#interface range fastethernet 0/20 - 24`	Moves to interface range config mode *20-24 TO BE ASSOCIATED WITH ETHERCHAN BUNDLE*
`3560Switch(config-if-range)#no ip address` *(CURRICULUM DOES NO SWPT)*	Ensures there are no IP addresses assigned on the interfaces
`3560Switch(config-if-range)#channel-protocol pagp`	Specifies the PAgP protocol to be used in this channel

3560Switch(config-if-range)#**channel-protocol lacp**	Specifies the LACP protocol to be used in this channel
3560Switch(config-if-range)#**channel-group 1 mode {desirable ¦ auto ¦ on ¦ passive ¦ active }**	Creates channel group 1 and assigns interfaces 20–24 as part of it. Use whichever mode is necessary, depending on your choice of protocol.
	NOTE: The channel group number must match the port channel number.

Verifying EtherChannel

Switch#**show running-config**	Displays list of what is currently running on the device
Switch#**show running-config interface fastethernet 0/12**	Displays interface fastethernet 0/12 information
Switch#**show interfaces fastethernet 0/12 etherchannel**	Displays L3 EtherChannel information
Switch#**show etherchannel**	Displays all EtherChannel information
Switch#**show etherchannel 1 port-channel**	Displays port channel information
Switch#**show etherchannel summary**	Displays a summary of EtherChannel information
Switch#**show pagp neighbor**	Shows PAgP neighbor information
Switch#**clear pagp 1 counters**	Clears PAgP channel group 1 information
Switch#**clear lacp 1 counters**	Clears LACP channel group 1 information
Switch(config)#**port-channel load-balance** *type*	Configures load balancing of method named *type*

	NOTE: The following methods are allowed when load balancing across a port channel: **dst-ip** Distribution is based on destination host IP address. **dst-mac** Distribution is based on the destination MAC address. Packets to the same destination are sent on the same port, but packets to different destinations are sent on different ports in the channel. **src-dst-ip** Distribution is based on source and destination host IP address. **src-dst-mac** Distribution is based on source and destination MAC address. **src-ip** Distribution is based on source IP address. **src-mac** Distribution is based on source MAC address. Packets from different hosts use different ports in the channel, but packets from the same host use the same port.
Switch#**show etherchannel load-balance**	Displays EtherChannel load-balancing information

Configuration Example: EtherChannel

Figure 3-2 shows the network topology for the configuration that follows, which shows how to configure EtherChannel using commands covered in this chapter.

Figure 3-2   *Network Topology for EtherChannel Configuration*

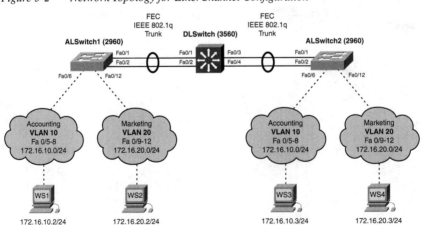

DLSwitch (3560)

`Switch>`**`enable`**	Moves to privileged mode
`Switch#`**`configure terminal`**	Moves to global config mode
`Switch(config)#`**`hostname DLSwitch`**	Sets host name
`DLSwitch(config)#`**`no ip domain-lookup`**	Turns off DNS queries so that spelling mistakes will not slow you down
`DLSwitch(config)#`**`vtp mode server`**	Changes the switch to VTP server mode
`DLSwitch(config)#`**`vtp domain testdomain`**	Configures the VTP domain name to testdomain
`DLSwitch(config)#`**`vlan 10`**	Creates VLAN 10 and enters VLAN-config mode
`DLSwitch(config-vlan)#`**`name Accounting`**	Assigns a name to the VLAN
`DLSwitch(config-vlan)#`**`exit`**	Returns to global config mode
`DLSwitch(config)#`**`vlan 20`**	Creates VLAN 20 and enters VLAN-config mode
`DLSwitch(config-vlan)#`**`name Marketing`**	Assigns a name to the VLAN
`DLSwitch(config-vlan)#`**`exit`**	Returns to global config mode
`DLSwitch(config)#`**`interface range fastethernet 0/1 - 4`**	Moves to interface range config mode
`DLSwitch(config-if)#`**`switchport trunk encapsulation dot1q`**	Specifies 802.1Q encapsulation on the trunk link
`DLSwitch(config-if)#`**`switchport mode trunk`**	Puts the interface into permanent trunking mode and negotiates to convert the link into a trunk link
`DLSwitch(config-if)#`**`exit`**	Returns to global config mode
`DLSwitch(config)#`**`interface range fastethernet 0/1 - 2`**	Moves to interface range config mode
`DLSwitch(config-if)#`**`channel-group 1 mode desirable`**	Creates channel group 1 and assigns interfaces 01–02 as part of it
`DLSwitch(config-if)#`**`exit`**	Moves to global config mode

DLSwitch(config)#`interface range fastethernet 0/3 - 4`	Moves to interface range config mode
DLSwitch(config-if)#`channel-group 2 mode desirable`	Creates channel group 2 and assigns interfaces 03–04 as part of it
DLSwitch(config-if)#`exit`	Moves to global config mode
DLSwitch(config)#`port-channel load-balance dst-mac`	Configures load balancing based on destination MAC address
DLSwitch(config)#`exit`	Moves to privileged mode
DLSwitch#`copy running-config startup-config`	Saves the configuration to NVRAM

ALSwitch1 (2960)

Switch>`enable`	Moves to privileged mode
Switch#`configure terminal`	Moves to global config mode
Switch(config)#`hostname ALSwitch1`	Sets host name
ALSwitch1(config)#`no ip domain-lookup`	Turns off DNS queries so that spelling mistakes will not slow you down
ALSwitch1(config)#`vtp mode client`	Changes the switch to VTP client mode
ALSwitch1(config)#`vtp domain testdomain`	Configures the VTP domain name to testdomain
ALSwitch1(config)#`interface range fastethernet 0/5 - 8`	Moves to interface range config mode
ALSwitch1(config-if-range)#`switchport mode access`	Sets ports 5–8 as access ports
ALSwitch1(config-if-range)#`switchport access vlan 10`	Assigns ports to VLAN 10
ALSwitch1(config-if-range)#`exit`	Moves to global config mode
ALSwitch1(config)#`interface range fastethernet 0/9 - 12`	Moves to interface range config mode
ALSwitch1(config-if-range)#`switchport mode access`	Sets ports 9–12 as access ports
ALSwitch1(config-if-range)#`switchport access vlan 20`	Assigns ports to VLAN 20

ALSwitch1(config-if-range)#exit	Moves to global config mode
ALSwitch1(config)#interface range fastethernet 0/1 - 2	Moves to interface range config mode
ALSwitch1(config-if-range)#switchport mode trunk	Puts the interface into permanent trunking mode and negotiates to convert the link into a trunk link
ALSwitch1(config-if-range)#channel-group 1 mode desirable	Creates Channel Group 1 and assigns interfaces 01–02 as part of it
ALSwitch1(config-if-range)#exit	Moves to global config mode
ALSwitch1(config)#exit	Moves to privileged mode
ALSwitch1#copy running-config startup-config	Saves the configuration to NVRAM

ALSwitch2 (2960)

Switch>enable	Moves to privileged mode
Switch#configure terminal	Moves to global config mode
Switch(config)#hostname ALSwitch2	Sets host name
ALSwitch2(config)#no ip domain-lookup	Turns off DNS queries so that spelling mistakes will not slow you down
ALSwitch2(config)#vtp mode client	Changes the switch to VTP client mode
ALSwitch2(config)#vtp domain testdomain	Configures the VTP domain name to testdomain
ALSwitch2(config)#interface range fastethernet 0/5 - 8	Moves to interface range config mode
ALSwitch2(config-if-range)#switchport mode access	Sets ports 5–8 as access ports
ALSwitch2(config-if-range)#switchport access vlan 10	Assigns ports to VLAN 10
ALSwitch2(config-if-range)#exit	Moves to global config mode

ALSwitch2(config)#**interface range fastethernet 0/9 - 12**	Moves to interface range config mode
ALSwitch2(config-if-range)#**switchport mode access**	Sets ports 9–12 as access ports
ALSwitch2(config-if-range)#**switchport access vlan 20**	Assigns ports to VLAN 20
ALSwitch2(config-if-range)#**exit**	Moves to global config mode
ALSwitch2(config)#**interface range fastethernet 0/1 - 2**	Moves to interface range config mode
ALSwitch2(config-if-range)#**switchport mode trunk**	Puts the interface into permanent trunking mode and negotiates to convert the link into a trunk link
ALSwitch2(config-if-range)#**channel-group 1 mode desirable**	Creates channel group 1 and assigns interfaces 01–02 as part of it.
ALSwitch2(config-if-range)#**exit**	Moves to global config mode
ALSwitch2(config)#**exit**	Moves to privileged mode
ALSwitch2#**copy running-config startup-config**	Saves the configuration to NVRAM

- SHOW PORT CAPABILITIES   CHECKS MODULE FOR ETHERCHANNEL FEATURE

- SHOW INTERFACE TRUNK

CHAPTER 4

Inter-VLAN Routing

This chapter provides information and commands concerning the following topics:

- Configuring Cisco Express Forwarding (CEF)
- Verifying CEF
- Troubleshooting CEF
- Inter-VLAN communication using an external router: router-on-a-stick
- Inter-VLAN communication tips
- Inter-VLAN communication on a multilayer switch through a switch virtual interface
  - Removing L2 switchport capability of a switch port
  - Configuring inter-VLAN communication

Configuring Cisco Express Forwarding

Switch(config)#**ip cef**	Enables standard CEF
Switch(config)#**ip cef distributed**	Enables distributed CEF (dCEF)
Switch(config)#**no ip cef**	Disables CEF globally
Switch(config)#**interface fastethernet 0/1**	Moves to interface configuration mode
Switch(config-if)#**ip route-cache cef**	Enables CEF on the interface

Verifying CEF

Switch#`show ip cef`	Displays entries in the Forwarding Information Base (FIB)
Switch#`show ip cef summary`	Displays a summary of the FIB
Switch#`show ip cef unresolved`	Displays unresolved FIB entries
Switch#`show ip cef fastethernet 0/1`	Displays the FIB entry for the specified interface
Switch#`show ip cef fastethernet 0/1 detail`	Displays detailed information about the FIB for the interface
Switch#`show interface fastethernet 0/1 ¦ begin L3`	Displays switching statistics for the interface beginning at the section for L3
Switch#`show interface gigabitethernet 1/1 ¦ include switched`	Displays switching statistics that show statistics for each layer
Switch#`show adjacency fastethernet 0/20 detail`	Displays the content of the information to be used during L2 encapsulation
Switch#`show cef drop`	Display packets that are dropped because adjacencies are incomplete or nonexistent
Switch#`show ip interface vlan10`	Verifies whether CEF is enabled on an interface

Troubleshooting CEF

Switch#`debug ip cef`	Displays debug information for CEF
Switch#`debug ip cef drop`	Displays debug information about dropped packets
Switch#`debug ip cef access-list x`	Displays information from specified access lists
Switch#`debug ip cef receive`	Displays packets that are not switched using information from the FIB but that are received and sent to the next switching layer
Switch#`debug ip cef events`	Displays general CEF events

`Switch#debug ip cef prefix-ipc`	Displays updates related to IP prefix information
`Switch#debug ip cef table`	Produces a table showing events related to the FIB table
`Switch#ping ip`	Performs an extended ping

Inter-VLAN Communication Using an External Router: Router-on-a-Stick

`Router(config)#interface fastethernet 0/0`	Moves to interface configuration mode
`Router(config-if)#duplex full`	Sets interface to full duplex
`Router(config-if)#no shutdown`	Enables interface
`Router(config-if)#interface fastethernet 0/0.1`	Creates subinterface 0/0.1 and moves to subinterface configuration mode
`Router(config-subif)#description Management VLAN 1`	(Optional) Sets locally significant descriptor of the subinterface
`Router(config-subif)#encapsulation dot1q 1 native`	Assigns VLAN 1 to this subinterface. VLAN 1 will be the native VLAN. This subinterface will use the 802.1Q trunking protocol.
`Router(config-subif)#ip address 192.168.1.1 255.255.255.0`	Assigns IP address and netmask
`Router(config-subif)#int fastethernet 0/0.10`	Creates subinterface 0/0.10 and moves to subinterface configuration mode
`Router(config-subif)#description Accounting VLAN 10`	(Optional) Sets locally significant descriptor of the subinterface
`Router(config-subif)#encapsulation dot1q 10`	Assigns VLAN 10 to this subinterface. This subinterface will use the 802.1Q trunking protocol.
`Router(config-subif)#ip address 192.168.10.1 255.255.255.0`	Assigns IP address and netmask
`Router(config-subif)#exit`	Returns to interface configuration mode
`Router(config-if)#exit`	Returns to global configuration mode
`Router(config)#`	

> **NOTE:** The subnets of the VLANs are directly connected to the router. Routing between these subnets does not require a dynamic routing protocol. In a more complex topology, these routes would need to either be advertised with whatever dynamic routing protocol is being used, or be redistributed into whatever dynamic routing protocol is being used.

> **NOTE:** Routes to the subnets associated with these VLANs will appear in the routing table as directly connected networks.

Inter-VLAN Communication Tips

- Although most routers support both Inter-Switch Link (ISL) and Dot1Q encapsulation some switch models only support Dot1Q, such as the 2950 and 2960 series.
- If you need to use ISL as your trunking protocol, use the command **encapsulation isl** *x*, where *x* is the number of the VLAN to be assigned to that subinterface.
- Recommended best practice is to use the same number of the VLAN number for the subinterface number. It is easier to troubleshoot VLAN 10 on subinterface fa0/0.10 than on fa0/0.2
- The native VLAN (usually VLAN 1) cannot be configured on a subinterface for Cisco IOS releases that are earlier than 12.1(3)T. Native VLAN IP addresses will therefore need to be configured on the physical interface. Other VLAN traffic will be configured on subinterfaces:

```
Router(config)#int fastethernet 0/0
Router(config-if)#encapsulation dot1q 1 native
Router(config-if)#ip address 192.168.1.1 255.255.255.0
Router(config-if)#int fastethernet 0/0.10
Router(config-subif)#encapsulation dot1q 10
Router(config-subif)#ip address 192.168.10.1 255.255.255.0
```

Inter-VLAN Communication on a Multilayer Switch Through a Switch Virtual Interface

Rather than using an external router to provide inter-VLAN communication, a multilayer switch can perform the same task through the use of a switched virtual interface (SVI).

Removing L2 Switchport Capability of a Switch Port

`3560Switch(config)#interface fastethernet 0/1`	Moves to interface configuration mode
`3560Switch(config-if)#no switchport`	Creates a Layer 3 port on the switch
	NOTE: The **no switchport** command can be used on physical ports only.

Configuring Inter-VLAN Communication

`3560Switch(config)#`**`interface vlan 1`**	Creates a virtual interface for VLAN 1 and enters interface configuration mode
`3550Switch(config-if)#`**`ip address 172.16.1.1 255.255.255.0`**	Assigns IP address and netmask
`3550Switch(config-if)#`**`no shutdown`**	Enables the interface
`3550Switch(config)#`**`interface vlan 10`**	Creates a virtual interface for VLAN 10 and enters interface configuration mode
`3550Switch(config-if)#`**`ip address 172.16.10.1 255.255.255.0`**	Assigns IP address and netmask
`3550Switch(config-if)#`**`no shutdown`**	Enables the interface
`3550Switch(config)#`**`interface vlan 20`**	Creates a virtual interface for VLAN 20 and enters interface configuration mode
`3550Switch(config-if)#`**`ip address 172.16.20.1 255.255.255.0`**	Assigns IP address and netmask
`3550Switch(config-if)#`**`no shutdown`**	Enables the interface
`3550Switch(config-if)#`**`exit`**	Returns to global configuration mode
`3550Switch(config)#`**`ip routing`**	Enables routing on the switch

NOTE: The subnets of the VLANs are directly connected to the switch. Routing between these subnets does not require a dynamic routing protocol. If the switch is to be connected to a router and remote communication is desired, a routing protocol must be enabled and networks advertised:

```
3560Switch(config)#router eigrp 1
3560Switch(config-router)#network 172.16.0.0
3560Switch(config-router)#exit
3560Switch(config)#
```

Configuration Example: Inter-VLAN Communication

Figure 4-1 shows the network topology for the configuration that follows, which shows how to configure inter-VLAN communication using commands covered in this chapter. Some commands used in this configuration are from previous chapters.

Figure 4-1   Network Topology for Inter-VLAN Communication Configuration

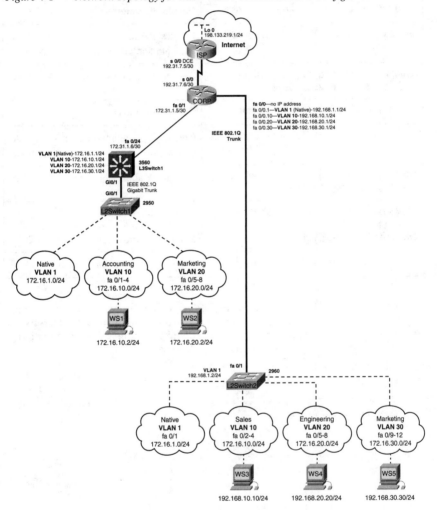

ISP Router

`Router>`**`enable`**	Moves to privileged mode
`Router>#`**`configure terminal`**	Moves to global config mode
`Router(config)#`**`hostname ISP`**	Sets host name
`ISP(config)#`**`interface loopback 0`**	Moves to interface configuration mode
`ISP(config-if)#`**`description simulated address representing remote website`**	Sets locally significant interface description
`ISP(config-if)#`**`ip address 198.133.219.1 255.255.255.0`**	Assigns IP address and netmask
`ISP(config-if)#`**`interface serial 0/0`**	Moves to interface configuration mode
`ISP(config-if)#`**`description WAN link to the Corporate Router`**	Sets locally significant interface description
`ISP(config-if)#`**`ip address 192.31.7.5 255.255.255.252`**	Assigns IP address and netmask
`ISP(config-if)#`**`clock rate 56000`**	Assigns a clock rate to the interface—DCE cable is plugged into this interface
`ISP(config-if)#`**`no shutdown`**	Enables the interface
`ISP(config-if)#`**`exit`**	Returns to global configuration mode
`ISP(config-if)#`**`router eigrp 10`**	Creates Enhanced Interior Gateway Routing Protocol (EIGRP) routing process 10
`ISP(config-router)#`**`network 198.133.219.0`**	Advertises directly connected networks (classful address only)
`ISP(config-router)#`**`network 192.31.7.0`**	Advertises directly connected networks (classful address only)
`ISP(config-router)#`**`no auto-summary`**	Disables auto summarization
`ISP(config-router)#`**`exit`**	Returns to global configuration mode
`ISP(config)#`**`exit`**	Returns to privileged mode
`ISP#`**`copy running-config startup-config`**	Saves the configuration to NVRAM

CORP Router

`Router>enable`	Moves to privileged mode
`Router>#configure terminal`	Moves to global configuration mode
`Router(config)#hostname CORP`	Sets host name
`CORP(config)#no ip domain-lookup`	Turns off Domain Name System (DNS) resolution to avoid wait time due to DNS lookup of spelling errors
`CORP(config)#interface serial 0/0`	Moves to interface configuration mode
`CORP(config-if)#description link to ISP`	Sets locally significant interface description
`CORP(config-if)#ip address 192.31.7.6 255.255.255.252`	Assigns IP address and netmask
`CORP(config-if)#no shutdown`	Enables interface
`CORP(config)#interface fastethernet 0/1`	Moves to interface configuration mode
`CORP(config-if)#description link to 3560 Switch`	Sets locally significant interface description
`CORP(config-if)#ip address 172.31.1.5 255.255.255.252`	Assigns IP address and netmask
`CORP(config-if)#no shutdown`	Enables interface
`CORP(config-if)#exit`	Returns to global configuration mode
`CORP(config)#interface fastethernet 0/0`	Enters interface configuration mode
`CORP(config-if)#duplex full`	Enables full-duplex operation to ensure trunking will take effect between here and L2Switch2
`CORP(config-if)#no shutdown`	Enables interface
`CORP(config-if)#interface fastethernet 0/0.1`	Creates a virtual subinterface and moves to subinterface configuration mode
`CORP(config-subif)#description Management VLAN 1 - Native VLAN`	Sets locally significant interface description

`CORP(config-subif)#`**`encapsulation dot1q 1`** **`native`**	Assigns VLAN 1 to this subinterface. VLAN 1 will be the native VLAN. This subinterface will use the 802.1Q trunking protocol.
`CORP(config-subif)#`**`ip address`** **`192.168.1.1 255.255.255.0`**	Assigns IP address and netmask
`CORP(config-subif)#`**`interface`** **`fastethernet 0/0.10`**	Creates a virtual subinterface and moves to subinterface configuration mode
`CORP(config-subif)#`**`description Sales`** **`VLAN 10`**	Sets locally significant interface description
`CORP(config-subif)#`**`encapsulation dot1q`** **`10`**	Assigns VLAN 10 to this subinterface. This subinterface will use the 802.1Q trunking protocol.
`CORP(config-subif)#`**`ip address`** **`192.168.10.1 255.255.255.0`**	Assigns IP address and netmask
`CORP(config-subif)#`**`interface`** **`fastethernet 0/0.20`**	Creates a virtual subinterface and moves to subinterface configuration mode
`CORP(config-subif)#`**`description`** **`Engineering VLAN 20`**	Sets locally significant interface description
`CORP(config-subif)#`**`encapsulation dot1q`** **`20`**	Assigns VLAN 20 to this subinterface. This subinterface will use the 802.1Q trunking protocol.
`CORP(config-subif)#`**`ip address`** **`192.168.20.1 255.255.255.0`**	Assigns IP address and netmask
`CORP(config-subif)#`**`interface`** **`fastethernet 0/0.30`**	Creates a virtual subinterface and moves to subinterface configuration mode
`CORP(config-subif)#`**`description`** **`Marketing VLAN 30`**	Sets locally significant interface description
`CORP(config-subif)#`**`encapsulation dot1q`** **`30`**	Assigns VLAN 30 to this subinterface. This subinterface will use the 802.1Q trunking protocol.
`CORP(config-subif)#`**`ip add 192.168.30.1`** **`255.255.255.0`**	Assigns IP address and netmask
`CORP(config-subif)#`**`exit`**	Returns to interface configuration mode

`CORP(config-if)#exit`	Returns to global configuration mode
`CORP(config)#router eigrp 10`	Creates EIGRP routing process 10 and moves to router configuration mode
`CORP(config-router)#network 192.168.1.0`	Advertises the 192.168.1.0 network
`CORP(config-router)#network 192.168.10.0`	Advertises the 192.168.10.0 network
`CORP(config-router)#network 192.168.20.0`	Advertises the 192.168.20.0 network
`CORP(config-router)#network 192.168.30.0`	Advertises the 192.168.30.0 network
`CORP(config-router)#network 172.31.0.0`	Advertises the 172.31.0.0 network
`CORP(config-router)#network 192.31.7.0`	Advertises the 192.31.7.0 network
`CORP(config-router)#no auto-summary`	Turns off automatic summarization at classful boundary
`CORP(config-router)#exit`	Returns to global configuration mode
`CORP(config)#exit`	Returns to privileged mode
`CORP#copy running-config startup-config`	Saves the configuration in NVRAM

L2Switch2 (Catalyst 2960)

`Switch>enable`	Moves to privileged mode
`Switch#configure terminal`	Moves to global configuration mode
`Switch(config)#hostname L2Switch2`	Sets host name
`L2Switch2(config)#no ip domain-lookup`	Turns off DNS resolution
`L2Switch2(config)#vlan 10`	Creates VLAN 10 and enters VLAN-configuration mode
`L2Switch2(config-vlan)#name Sales`	Assigns a name to the VLAN
`L2Switch2(config-vlan)#exit`	Returns to global configuration mode

`L2Switch2(config)#`**`vlan 20`**	Creates VLAN 20 and enters VLAN-configuration mode
`L2Switch2(config-vlan)#`**`name Engineering`**	Assigns a name to the VLAN
`L2Switch2(config-vlan)#`**`vlan 30`**	Creates VLAN 30 and enters VLAN-configuration mode. Note that you do not have to exit back to global configuration mode to execute this command.
`L2Switch2(config-vlan)#`**`name Marketing`**	Assigns a name to the VLAN
`L2Switch2(config-vlan)#`**`exit`**	Returns to global configuration mode
`L2Switch2(config)#`**`interface range fastethernet 0/2 - 4`**	Enables you to set the same configuration parameters on multiple ports at the same time
`L2Switch2(config-if-range)#`**`switchport mode access`**	Sets ports 2–4 as access ports
`L2Switch2(config-if-range)#`**`switchport access vlan 10`**	Assigns ports 2–4 to VLAN 10
`L2Switch2(config-if-range)#`**`interface range fastethernet 0/5 - 8`**	Enables you to set the same configuration parameters on multiple ports at the same time
`L2Switch2(config-if-range)#`**`switchport mode access`**	Sets ports 5–8 as access ports
`L2Switch2(config-if-range)#`**`switchport access vlan 20`**	Assigns ports 5–8 to VLAN 20
`L2Switch2(config-if-range)#`**`interface range fastethernet 0/9 - 12`**	Enables you to set the same configuration parameters on multiple ports at the same time
`L2Switch2(config-if-range)#`**`switchport mode access`**	Sets ports 9–12 as access ports
`L2Switch2(config-if-range)#`**`switchport access vlan 30`**	Assigns ports 9–12 to VLAN 30
`L2Switch2(config-if-range)#`**`exit`**	Returns to global configuration mode
`L2Switch2(config)#`**`int fastethernet 0/1`**	Moves to interface configuration mode
`L2Switch2(config)#`**`description Trunk Link to CORP Router`**	Sets locally significant interface description

L2Switch2(config-if)#**switchport mode trunk**	Puts the interface into trunking mode and negotiates to convert the link into a trunk link
L2Switch2(config-if)#**exit**	Returns to global configuration mode
L2Switch2(config)#**interface vlan 1**	Creates virtual interface for VLAN 1 and enters interface configuration mode
L2Switch2(config-if)#**ip address 192.168.1.2 255.255.255.0**	Assigns IP address and netmask
L2Switch2(config-if)#**no shutdown**	Enables interface
L2Switch2(config-if)#**exit**	Returns to global configuration mode
L2Switch2(config)#**ip default-gateway 192.168.1.1**	Assigns default gateway address
L2Switch2(config)#**exit**	Returns to privileged mode
L2Switch2#**copy running-config startup-config**	Saves the configuration in NVRAM

L3Switch1 (Catalyst 3560)

Switch>**enable**	Moves to privileged mode
Switch#**configure terminal**	Moves to global configuration mode
Switch(config)#**hostname L3Switch1**	Sets Hostname
L3Switch1(config)#**no ip domain-lookup**	Turns off DNS queries so that spelling mistakes will not slow you down
L3Switch1(config)#**vtp mode sever**	Changes the switch to VTP server mode
L3Switch1(config)#**vtp domain testdomain**	Configures the VTP domain name to **testdomain**
L3Switch1(config)#**vlan 10**	Creates VLAN 10 and enters VLAN-configuration mode
L3Switch1(config-vlan)#**name Accounting**	Assigns a name to the VLAN
L3Switch1(config-vlan)#**exit**	Returns to global configuration mode

`L3Switch1(config)#`**`vlan 20`**	Creates VLAN 20 and enters VLAN-configuration mode
`L3Switch1(config-vlan)#`**`name Marketing`**	Assigns a name to the VLAN
`L3Switch1(config-vlan)#`**`exit`**	Returns to global configuration mode
`L3Switch1(config)#`**`interface gigabitethernet 0/1`**	Moves to interface configuration mode
`L3Switch1(config-if)#`**`switchport trunk encapsulation dot1q`**	Specifies 802.1Q encapsulation on the trunk link
`L3Switch1(config-if)#`**`switchport mode trunk`**	Puts the interface into trunking mode and negotiates to convert the link into a trunk link
`L3Switch1(config-if)#`**`exit`**	Returns to global configuration mode
`L3Switch1(config)#`**`ip routing`**	Enables IP routing on this device
`L3Switch1(config)#`**`interface vlan 1`**	Creates virtual interface for VLAN 1 and enters interface configuration mode
`L3Switch1(config-if)#`**`ip address 172.16.1.1 255.255.255.0`**	Assigns IP address and netmask
`L3Switch1(config-if)#`**`no shutdown`**	Enables interface
`L3Switch1(config-if)#`**`interface vlan 10`**	Creates virtual interface for VLAN 10 and enters interface configuration mode
`L3Switch1(config-if)#`**`ip address 172.16.10.1 255.255.255.0`**	Assigns IP address and mask
`L3Switch1(config-if)#`**`no shutdown`**	Enables interface
`L3Switch1(config-if)#`**`interface vlan 20`**	Creates virtual interface for VLAN 20 and enters interface configuration mode
`L3Switch1(config-if)#`**`ip address 172.16.20.1 255.255.255.0`**	Assigns IP address and mask
`L3Switch1(config-if)#`**`no shutdown`**	Enables interface
`L3Switch1(config-if)#`**`exit`**	Returns to global configuration mode

`L3Switch1(config)#interface fastethernet 0/24`	Enters interface configuration mode
`L3Switch1(config-if)#no switchport`	Creates a Layer 3 port on the switch
`L3Switch1(config-if)#ip address 172.31.1.6 255.255.255.252`	Assigns IP address and netmask
`L3Switch1(config-if)#exit`	Returns to global configuration mode
`L3Switch1(config)#router eigrp 10`	Creates EIGRP routing process 10 and moves to router config mode
`L3Switch1(config-router)#network 172.16.0.0`	Advertises the 172.16.0.0 classful network
`L3Switch1(config-router)#network 172.31.0.0`	Advertises the 172.31.0.0 classful network
`L3Switch1(config-router)#no auto-summary`	Turns off automatic summarization at classful boundary
`L3Switch1(config-router)#exit`	Applies changes and returns to global configuration mode
`L3Switch1(config)#exit`	Returns to privileged mode
`L3Switch1#copy running-config startup-config`	Saves configuration in NVRAM

L2Switch1 (Catalyst 2960)

`Switch>enable`	Moves to privileged mode
`Switch#configure terminal`	Moves to global configuration mode
`Switch(config)#hostname L2Switch1`	Sets host name
`L2Switch1(config)#no ip domain-lookup`	Turns off DNS queries so that spelling mistakes will not slow you down
`L2Switch1(config)#vtp domain testdomain`	Configures the VTP domain name to testdomain
`L2Switch1(config)#vtp mode client`	Changes the switch to VTP client mode

L2Switch1(config)#**interface range fastethernet 0/1 - 4**	Enables you to set the same configuration parameters on multiple ports at the same time
L2Switch1(config-if-range)#**switchport mode access**	Sets ports 1–4 as access ports
L2Switch1(config-if-range)#**switchport access vlan 10**	Assigns ports 1–4 to VLAN 10
L2Switch1(config-if-range)#**interface range fastethernet 0/5 - 8**	Enables you to set the same configuration parameters on multiple ports at the same time
L2Switch1(config-if-range)#**switchport mode access**	Sets ports 5–8 as access ports
L2Switch1(config-if-range)#**switchport access vlan 20**	Assigns ports 5–8 to VLAN 20
L2Switch1(config-if-range)#**exit**	Returns to global configuration mode
L2Switch1(config)#**interface gigabitethernet 0/1**	Moves to interface configuration mode
L2Switch1(config-if)#**switchport mode trunk**	Puts the interface into trunking mode and negotiates to convert the link into a trunk link
L2Switch1(config-if)#**exit**	Returns to global configuration mode
L2Switch1(config)#**interface vlan 1**	Creates virtual interface for VLAN 1 and enters interface configuration mode
L2Switch1(config-if)#**ip address 172.16.1.2 255.255.255.0**	Assigns IP address and netmask
L2Switch1(config-if)#**no shutdown**	Enables interface
L2Switch1(config-if)#**exit**	Returns to global configuration mode
L2Switch1(config)#**ip default-gateway 172.16.1.1**	Assigns default gateway address
L2Switch1(config)#**exit**	Returns to privileged mode
L2Switch1#**copy running-config startup-config**	Saves the configuration in NVRAM

CHAPTER 5

High Availability

This chapter provides information and commands concerning the following topics:

- Hot Standby Routing Protocol (HSRP)
  — Configuring HSRP
  — Verifying HSRP
  — HSRP optimization options
  — Debugging HSRP
- Virtual Router Redundancy Protocol (VRRP)
  — Configuring VRRP
  — Verifying VRRP
  — Debugging VRRP
- Gateway Load Balancing Protocol (GLBP)
  — Configuring GLBP
  — Verifying GLBP
  — Debugging GLBP

Hot Standby Routing Protocol

The Hot Standby Router Protocol (HSRP) provides network redundancy for IP networks, ensuring that user traffic immediately and transparently recovers from first-hop failures in network edge devices or access circuits.

Configuring HSRP

`Router(config)#interface fastethernet 0/0`	Moves to interface configuration mode
`Router(config-if)#ip address 172.16.0.10 255.255.255.0`	Assigns IP address and netmask
`Router(config-if)#standby 1 ip 172.16.0.1`	Activates HSRP group 1 on the interface and creates a virtual IP address of 172.16.0.1 for use in HSRP
	NOTE: The group number can be from 0 to 255. The default is 0.
`Router(config-if)#standby 1 priority 120`	Assigns a priority value of 120 to standby group 1

	NOTE: The priority value can be from 1 to 255. The default is 100. A higher priority will result in that router being elected the active router. If the priorities of all routers in the group are equal, the router with the *highest IP address* becomes the active router.

Verifying HSRP

Router#show running-config	Displays what is currently running on the router
Router#show standby	Displays HSRP information
Router#show standby brief	Displays a single-line output summary of each standby group
Router#show standby 1	Displays HSRP group 1 information
Router#show standby fastethernet 0/0	Displays HSRP information for the specified interface
Router#show standby fastethernet 0/0 brief	Displays a summary of HSRP for the specified interface
Router#show standby fastethernet 0/0 1	Displays HSRP group 1 information for the specified interface

HSRP Optimization Options

There are options available that make it possible to optimize HSRP operation in the campus network. The next three sections explain three of these options: standby preempt, message timers, and interface tracking.

Preempt

Router(config)#interface fastethernet 0/0	Moves to interface configuration mode
Router(config-if)#standby 1 preempt	This router will preempt, or take control of, the active router if the local priority is higher than the active router

`Router(config-if)#standby 1 preempt delay minimum 180`	Causes the local router to postpone taking over as the active router for 180 seconds since that router was last restarted
`Router(config-if)#standby 1 preempt delay reload`	Allows for preemption to occur only after a router reloads
`Router(config-if)#no standby 1 preempt delay reload`	Disables the preemption delay, but preemption itself is still enabled. Use the **no standby** x **preempt** command to eliminate preemption
	NOTE: If the **preempt** argument is not configured, the local router assumes control as the active router only if the local router receives information indicating that there is no router currently in the active state.

HSRP Message Timers

`Router(config)#interface fastethernet 0/0`	Moves to interface config mode
`Router(config-if)#standby 1 timers 5 15`	Sets the hello timer to 5 seconds and sets the hold timer to 15 seconds
	NOTE: The hold timer is normally set to be greater than or equal to 3 times the hello timer.
	NOTE: The hello timer can be from 1 to 254; the default is 3. The hold timer can be from 1 to 255; the default is 10. The default unit of time is seconds.
`Router(config-if)#standby 1 timers msec 200 msec 600`	Sets the hello timer to 200 milliseconds and sets the hold timer to 600 milliseconds
	NOTE: If the **msec** argument is used, the timers can be an integer from 15 to 999.

Interface Tracking

`Router(config)#interface fastethernet 0/0`	Moves to interface configuration mode
`Router(config-if)#standby 1 track serial 0/0 25`	HSRP will track the availability of interface serial 0/0. If serial 0/0 goes down, the priority of the router in group 1 will be decremented by 25.
	NOTE: The default value of the **track** argument is 10.
	TIP: The **track** argument does not assign a new priority if the tracked interface goes down. The **track** argument assigns a value that the priority will be decreased if the tracked interface goes down. Therefore, if you are tracking serial 0/0 with a track value of 25—**standby 1 track serial 0/0 25**—and serial 0/0 goes down, the priority will be decreased by 25; assuming a default priority of 100, the new priority will now be 75.

Debugging HSRP

`Router#debug standby`	Displays all HSRP debugging information, including state changes and transmission/reception of HSRP packets
`Router#debug standby errors`	Displays HSRP error messages
`Router#debug standby events`	Displays HSRP event messages
`Router#debug standby events terse`	Displays all HSRP events except for hellos and advertisements
`Router#debug standby events track`	Displays all HSRP tracking events
`Router#debug standby packets`	Displays HSRP packet messages
`Router#debug standby terse`	Displays all HSRP errors, events, and packets, except for hellos and advertisements

Virtual Router Redundancy Protocol

NOTE: HSRP is Cisco proprietary. The Virtual Router Redundancy Protocol (VRRP) is an IEEE standard.

VRRP is an election protocol that dynamically assigns responsibility for one or more virtual routers to the VRRP routers on a LAN, allowing several routers on a multiaccess link to use the same virtual IP address. A VRRP router is configured to run VRRP in conjunction with one or more other routers attached.

Configuring VRRP

`Router(config)#interface fastethernet 0/0`	Moves to interface config mode
`Router(config-if)#ip address 172.16.100.5 255.255.255.0`	Assigns IP address and netmask
`Router(config-if)#vrrp 10 ip 172.16.100.1`	Enables VRRP for group 10 on this interface with a virtual address of 172.16.100.1. The group number can be from 1 to 255.
`Router(config-if)#vrrp 10 description Engineering Group`	Assigns a text description to the group
`Router(config-if)#vrrp 10 priority 110`	Sets the priority level for this router. The range is from 1 to 254. The default is 100.
`Router(config-if)#vrrp 10 preempt`	This router will preempt, or take over, as the virtual router master for group 10 if it has a higher priority than the current virtual router master
`Router(config-if)#vrrp 10 preempt delay minimum 60`	This router will preempt, but only after a delay of 60 seconds
	NOTE: The default delay period is 0 seconds.
`Router(config-if)#vrrp 10 timers advertise 15`	Configures the interval between successful advertisements by the virtual router master
	NOTE: The default interval value is 1 second.
	NOTE: All routers in a VRRP group must use the same timer values. If routers have different timer values set, the VRRP group will not communicate with each other.
	NOTE: The range of the advertisement timer is 1 to 255 seconds. If you use the **msec** argument, you change the timer to measure in milliseconds. The range in milliseconds is 50 to 999.

Router(config-if)#vrrp 10 timers learn	Configures the router, when acting as a virtual router backup, to learn the advertisement interval used by the virtual router master
Router(config-if)#vrrp 10 shutdown	Disables VRRP on the interface, but configuration is still retained
Router(config-if)#no vrrp 10 shutdown	Reenables the VRRP group using the previous configuration

Verifying VRRP

Router#show running-config	Displays contents of dynamic RAM
Router#show vrrp	Displays VRRP information
Router#show vrrp brief	Displays a brief status of all VRRP groups
Router#show vrrp 10	Displays detailed information about VRRP group 10
Router#show vrrp interface fastethernet 0/0	Displays information about VRRP as enabled on interface fastethernet 0/0
Router#show vrrp interface fastethernet 0/0 brief	Displays a brief summary about VRRP on interface fastethernet 0/0

Debugging VRRP

Router#debug vrrp all	Displays all VRRP messages
Router#debug vrrp error	Displays all VRRP error messages
Router#debug vrrp events	Displays all VRRP event messages
Router#debug vrrp packets	Displays messages about packets sent and received
Router#debug vrrp state	Displays messages about state transitions

Gateway Load Balancing Protocol

Gateway Load Balancing Protocol (GLBP) protects data traffic from a failed router or circuit, like HSRP and VRRP, while allowing packet load sharing between a group of redundant routers.

Configuring GLBP

`Router(config)#interface fastethernet 0/0`	Moves to interface config mode
`Router(config-if)#ip address 172.16.100.5 255.255.255.0`	Assigns IP address and netmask
`Router(config-if)#glbp 10 ip 172.16.100.1`	Enables GLBP for group 10 on this interface with a virtual address of 172.16.100.1. The range of group numbers is from 0 to 1023.
`Router(config-if)#glbp 10 preempt`	Configures the router to preempt, or take over, as the active virtual gateway (AVG) for group 10 if this router has a higher priority than the current AVG
`Router(config-if)#glbp 10 preempt delay minimum 60`	Configures the router to preempt, or take over, as AVG for group 10 if this router has a higher priority than the current active virtual forwarder (AVF) after a delay of 60 seconds
`Router(config-if)#glbp 10 forwarder preempt`	Configures the router to preempt, or take over, as AVF for group 10 if this router has a higher priority than the current AVF. This command is enabled by default with a delay of 30 seconds.
`Router(config-if)#glbp 10 forwarder preempt delay minimum 60`	Configures the router to preempt, or take over, as AVF for group 10 if this router has a higher priority than the current AVF after a delay of 60 seconds

	NOTE: Members of a GLBP group elect one gateway to be the AVG for that group. Other group members provide backup for the AVG in the event that the AVG becomes unavailable. The AVG assigns a virtual MAC address to each member of the GLBP group. Each gateway assumes responsibility for forwarding packets sent to the virtual MAC address assigned to it by the AVG. These gateways are known as AVFs for their virtual MAC address. Virtual forwarder redundancy is similar to virtual gateway redundancy with an AVF. If the AVF fails, one of the secondary virtual forwarders in the listen state assumes responsibility for the virtual MAC address.
`Router(config-if)#`**`glbp 10 priority 150`**	Sets the priority level of the router
	NOTE: The range of the **priority** argument is 1 to 255. The default priority of GLBP is 100. A higher priority number is preferred.
`Router(config-if)#`**`glbp 10 timers 5 15`**	Configures the hello timer to be set to 5 seconds and the hold timer to be 15 seconds
`Router(config-if)#`**`glbp 10 timers msec 20200 msec 60600`**	Configures the hello timer to be 20,200 milliseconds and the hold timer to be 60,600 milliseconds.
	NOTE: The default hello timer is 3 seconds. The range of the hello timer interval is 1 to 60 seconds. If the **msec** argument is used, the timer will be measured in milliseconds, with a range of 50 to 60000.

	NOTE: The default hold timer is 10 seconds. The range of the hold timer is 19 to 180 seconds. If the **msec** argument is used, the timer will be measured in milliseconds, with a range of 18020 to 180000.
	The hello timer measures the interval between successive hello packets sent by the AVG in a GLBP group. The **holdtime** argument specifies the interval before the virtual gateway and the virtual forwarder information in the hello packet is considered invalid. It is recommended that unless you are extremely familiar with your network design and with the mechanisms of GLBP that you do not change the timers. To reset the timers back to their default values, use the **no glbp** *x* **timers** command, where *x* is the GLBP group number.
`Router(config-if)#`**`glbp 10 load-`** **`balancing host-dependent`**	Specifies that GLBP will load balance using the host-dependent method
`Router(config-if)#`**`glbp 10 load-`** **`balancing weighted`**	Specifies that GLBP will load balance using the weighted method
`Router(config-if)#`**`glbp 10 weighting 80`**	Assigns a maximum weighting value for this interface for load-balancing purposes. The value can be from 1 to 254.
`Router(config-if)#`**`glbp 10 load`** **`balancing round robin`**	Specifies that GLBP will load balance using the round-robin method

NOTE: There are three different types of load balancing in GLBP:

- **Host-dependent** uses the MAC address of a host to determine which VF MAC address the host is directed toward. This is used with stateful Network Address Translation (NAT) because NAT requires each host to be returned to the same virtual MAC address each time it sends an ARP request for the virtual IP address. It is not recommended for situations where there are a small number of end hosts (fewer than 20).

- **Weighted** allows for GLBP to place a weight on each device when calculating the amount of load sharing. For example, if there are two routers in the group, and router A has twice the forwarding capacity of router B, the weighting value should be configured to be double the amount of router B. To assign a weighting value, use the **glbp** x **weighting** y interface configuration command, where x is the GLBP group number, and y is the weighting value, a number from 1 to 254.
- **Round-robin** load balancing occurs when each VF MAC address is used sequentially in ARP replies for the virtual IP address. Round robin is suitable for any number of end hosts.

If no load balancing is used with GLBP, GLBP will operate in an identical manner to HSRP, where the AVG will only respond to ARP requests with its own VF MAC address, and all traffic will be directed to the AVG.

Verifying GLBP

Router#`show running-config`	Displays contents of dynamic RAM
Router#`show glbp`	Displays GLBP information
Router#`show glbp brief`	Displays a brief status of all GLBP groups
Router#`show glbp 10`	Displays information about GLBP group 10
Router#`show glbp fastethernet 0/0`	Displays GLBP information on interface fastethernet 0/0
Router#`show glbp fastethernet 0/0 10`	Displays GLBP group 10 information on interface fastethernet 0/0

Debugging GLBP

Router#`debug condition glbp`	Displays GLBP condition messages
Router#`debug glbp errors`	Displays all GLBP error messages
Router#`debug glbp events`	Displays all GLBP event messages
Router#`debug glbp packets`	Displays messages about packets sent and received
Router#`debug glbp terse`	Displays a limited range of debugging messages

Configuration Example: HSRP

Figure 5-1 shows the network topology for the configuration that follows, which shows how to configure HSRP using the commands covered in this chapter. Note that only the commands specific to HSRP are shown in this example.

Figure 5-1   Network Topology for HSRP Configuration Example

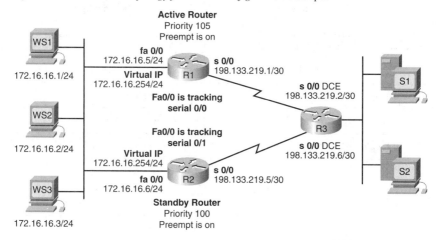

Router 1

`Router>`**`enable`**	Moves to privileged mode
`Router#`**`configure terminal`**	Moves to global configuration mode
`Router(config)#`**`hostname R1`**	Sets router name to R1
`R1(config)#`**`interface fastethernet 0/0`**	Moves to interface config mode
`R1(config-if)#`**`ip address 172.16.16.5 255.255.255.0`**	Assigns IP address and netmask
`R1(config-if)#`**`standby 1 ip 172.16.16.254`**	Activates HSRP group 1 on the interface and creates a virtual IP address of 172.16.6.254
`R1(config-if)#`**`standby 1 priority 105`**	Assigns a priority value of 105 to standby group 1
`R1(config-if)#`**`standby 1 preempt`**	This router will preempt, or take control of, the active router if the local priority is higher than the active router

`R1(config-if)#standby 1 track serial 0/0`	HSRP will track the availability of interface serial 0/0. If serial 0/0 goes down, the router priority will be decremented by the default 10.
`R1(config-if)#no shutdown`	Enables the interface
`R1(config-if)#interface serial 0/0`	Moves to interface config mode
`R1(config-if)#ip address 198.133.219.1 255.255.255.252`	Assigns IP address and netmask
`R1(config-if)#no shutdown`	Enables the interface
`R1(config-if)#exit`	Returns to global config mode
`R1(config)#exit`	Returns to privileged mode
`R1#copy running-config startup-config`	Saves the configuration to NVRAM

Router 2

`Router>enable`	Moves to privileged mode
`Router#configure terminal`	Moves to global config mode
`Router(config)#hostname R2`	Sets router name to R2
`R2(config)#interface fastethernet 0/0`	Moves to interface config mode
`R2(config-if)#ip address 172.16.16.6 255.255.255.0`	Assigns IP address and netmask
`R2(config-if)#standby 1 ip 171.16.16.254`	Activates HSRP group 1 on the interface and creates a virtual IP address of 172.16.6.254
`R2(config-if)#standby 1 priority 100`	Assigns a priority value of 100 to standby group 1
`R2(config-if)#standby 1 preempt`	This router will preempt, or take control of, the active router if the local priority is higher than the active router
`R2(config-if)#standby 1 track serial 0/1`	HSRP will track the availability of interface serial 0/1. If S0/1 goes down, the router priority will be decremented by the default 10.

`R2(config-if)#no shutdown`	Enables the interface
`R2(config-if)#interface serial 0/1`	Moves to interface config mode
`R2(config-if)#ip address` `198.133.219.5 255.255.255.252`	Assigns IP address and netmask
`R2(config-if)#no shutdown`	Enables the interface
`R2(config-if)#exit`	Returns to global config mode
`R2(config)#exit`	Returns to privileged mode
`R2#copy running-config startup-config`	Saves the configuration to NVRAM

Configuration Example: GLBP

Figure 5-2 shows the network topology for the configuration that follows, which shows how to configure GLBP using commands covered in this chapter. Note that only the commands specific to GLBP are shown in this example.

Figure 5-2   Network Topology for GLBP Configuration Example

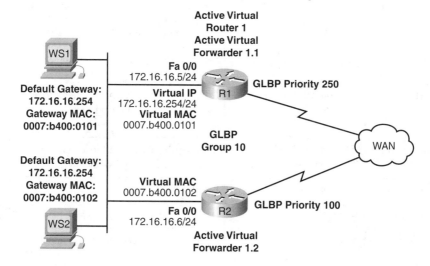

R1 is the AVG for a GLBP group and is responsible for the virtual IP address 10.21.8.10. R1 is also an AVF for the virtual MAC address 0007.b400.0101. R1 is a member of the same GLBP group and is designated as the AVF for the virtual MAC address 0007.b400.0102.

WS1 has a default gateway IP address of 10.21.8.10 and a gateway MAC address of 0007.b400.0101.

WS2 shares the same default gateway IP address but receives the gateway MAC address 0007.b400.0102 because R2 is sharing the traffic load with R1.

Router 1

`Router>`**`enable`**	Moves to privileged mode
`Router#`**`configure terminal`**	Moves to global config mode
`Router(config)#`**`hostname R1`**	Assigns router name
`R1(config)#`**`interface fastethernet 0/0`**	Moves to interface config mode
`R1(config-if)#`**`ip address 172.16.16.5 255.255.255.0`**	Assigns IP address and netmask
`R1(config-if)#`**`glbp 10 ip 172.16.16.254`**	Enables GLBP for group 10 on this interface with a virtual address of 172.16.16.254
`R1(config-if)#`**`glbp 10 preempt`**	Configures the router to preempt, or take over, as AVG for group 10 if this router has a higher priority than the current AVG
`R1(config-if)#`**`glbp 10 priority 250`**	Sets the priority level of the router
`R1(config-if)#`**`glbp 10 timers 5 18`**	Configures the hello timer to be set to 5 seconds and the hold timer to be 18 seconds
`R1(config-if)#`**`glbp 10 load-balancing host-dependent`**	Specifies that GLBP will load balance using the host-dependent method
`R1(config-if)#`**`no shutdown`**	Enables the interface
`R1(config-if)#`**`exit`**	Returns to global config mode
`R1(config)#`**`exit`**	Returns to privileged mode
`R1#`**`copy running-config startup-config`**	Saves the configuration to NVRAM

Router 2

`Router>`**`enable`**	Moves to privileged mode
`Router#`**`configure terminal`**	Moves to global config mode
`Router(config)#`**`hostname R2`**	Assigns router name
`R2(config)#`**`interface fastethernet 0/0`**	Moves to interface config mode
`R2(config-if)#`**`ip address 172.16.16.6 255.255.255.0`**	Assigns IP address and netmask
`R2(config-if)#`**`glpb 10 ip 172.16.16.254`**	Enables GLBP for group 10 on this interface with a virtual address of 172.16.16.254
`R2(config-if)#`**`glbp 10 preempt`**	Configures the router to preempt, or take over, as AVG for group 10 if this router has a higher priority than the current AVG
`R2(config-if)#`**`glbp 10 priority 100`**	Sets the priority level of the router. The default setting is 100.
`R2(config-if)#`**`glbp 10 timers 5 18`**	Configures the hello timer to be set to 5 seconds and the hold timer to be 18 seconds
`R1(config-if)#`**`glbp 10 load-balancing host-dependent`**	Specifies that GLBP will load balance using the host-dependent method
`R2(config-if)#`**`no shutdown`**	Enables the interface
`R2(config-if)#`**`exit`**	Returns to global config mode
`R2(config)#`**`exit`**	Returns to privileged mode
`R2#`**`copy running-config startup-config`**	Saves the configuration to NVRAM

CHAPTER 6

Wireless Client Access

This chapter provides information and commands concerning the following topics:

- **Configuration Example: 4402 WLAN Controller Using the Configuration Wizard**—Covers the tasks and command-line interface (CLI) commands to
    - Assign a host name to the wireless LAN controller (WLC)
    - Enable Telnet access to the WLC
    - Enable HTTP access to the WLC
    - Set a timeout of the CLI
    - Save a configuration
    - Verify the configuration
- **Configuration Example: 4402 WLAN Controller Using the Web Interface**—Covers the same tasks as listed above in the CLI, but with the graphical user interface (GUI)
- **Configuration Example: Configuring a 3560 Switch to Support WLANs and APs**—Covers how to configure a 3560 switch to support wireless LANs (WLAN) and access points (AP)
- **Configuration Example: Configuring a Wireless Client**—Covers how to configure a Cisco Aironet wireless client adapter

Configuration Example: 4402 WLAN Controller Using the Configuration Wizard

NOTE: In the Wireless LAN Controller (WLC) Configuration Wizard, all available options appear in brackets after each parameter. The default value appears in all uppercase letters.

Commands are case sensitive.

```
     .o88b. d888888b .d8888.  .o88b.  .d88b.
   d8P  Y8  `88'  88'  YP d8P  Y8 .8P  Y8.
   8P        88  `8bo.  8P        88     88
   8b        88  `Y8b. 8b        88     88
   Y8b  d8 .88.   db   8D Y8b  d8 `8b  d8'
    `Y88P' Y888888P `8888Y'  `Y88P'  `Y88P'
   Model AIR-WLC4402-12-K9   S/N: XXXXXXXXXXX
Net:
 PHY DEVICE  : Found Intel LXT971A PHY at 0x01
FEC ETHERNET
IDE:   Bus 0: OK
 Device 0: Model: STI Flash 7.4.0 Firm: 01.25.06
Ser#: XXXXXXXXXXX
               Type: Removable Hard Disk
               Capacity: 245.0 MB = 0.2 GB (501760
x 512)
 Device 1: not available
```

`Booting Primary Image...` `Press <ESC> now for additional boot options...`	Select **1** to continue to boot the primary image—this is the default choice
`***** External Console Active *****`	Select **2** to boot the backup image (image used before the last software upgrade)
	Select **3** for manual upgrade of image files
`      Boot Options`	Select **4** to set the backup image as the primary image

`Please choose an option from below:`
 	
`1. Run primary image (version 4.0.179.8)`
`(active)`
`2. Run backup image (version 4.0.179.8)`
`3. Manually update images`
`4. Change active boot image`
`5. Clear Configuration`
 	
`Please enter your choice: 1`
 	
`Detecting Hardware . . .`
	Select **5** to set the configuration back to factory default and start the CLI Setup Wizard using the current software

	NOTE: Option 3 is for recovery only. Do not select this option unless you have the required files and are instructed to do so by the Cisco Technical Assistance Center (TAC).
`<OUTPUT CUT>`	
`Welcome to the Cisco Wizard Configuration Tool` `Use the '-' character to backup`	Press the hyphen key if you ever need to return to the previous command line
`System Name [Cisco_xx:xx:xx]:`	Enters the system name for the controller. Length is up to 32 ASCII characters. If no name is entered, a default of Cisco Controller is used.
`Enter Administrative User Name (24 characters max):` **`cisco`** `Enter Administrative Password (24 characters max):` **`password`**	Assigns the administrative username and password. The default username and password are admin and admin.
`Service Interface IP Address Configuration [none][DHCP]:`**`DHCP`**	Enter **DHCP** if you want the controller's service-port interface to obtain its IP address from a DHCP server. Enter **none** if you want to set one statically or if you don't want to use the service port.

	NOTE: The service-port interface controls communications through the service port. Its IP address must be on a different subnet from the management and AP-manager interfaces. This enables you to manage the controller directly or through a dedicated management network to ensure service access during network downtime.
	NOTE: If you do not want to use the service port, enter **0.0.0.0** for the IP address and subnet mask. If you want to statically assign an address and mask, do so on the next two lines when prompted.
`Enable Link Aggregation (LAG) [yes][NO]:`	Enables link aggregation, if desired
`Management Interface IP Address: `**`172.16.1.100`** `Management Interface Netmask: `**`255.255.255.0`** `Management Interface Default Router: `**`172.16.1.1`** `Management Interface VLAN Identifier (0 = untagged): `**`0`** `Management Interface Port Num [1 to 2]: `**`1`** `Management Interface DHCP Server IP Address:` **`172.16.1.1`**	Assigns IP address, netmask, default router IP address, optional VLAN identifier of the management interface, and port number of the management interface Assigns the IP address of the DHCP server that will assign addresses to the management interface and service-port interface
	NOTE: The VLAN identifier should be set to match the switch interface configuration.

	NOTE: The management interface is the default interface for in-band management of the controller and connectivity to enterprise services such as an authentication, authorization, and accounting (AAA) server.
`AP Transport Mode [layer2][LAYER3]:`	Sets AP transport layer

`AP Manager Interface IP Address: `**`172.16.100.100`**
 	
`AP Manager Interface Netmask: `**`255.255.255.0`**
 	
`AP Manager Interface Default Router:`
`172.16.100.1`
 	
`AP Manager Interface VLAN Identifier (0 =`
`untagged): `**`100`**
 	
`AP Manager Interface Port Num [1 to 2]: `**`1`**
 	
`AP Manager Interface DHCP Server (172.16.1.1):`
`172.16.100.1`
	Assigns IP address, netmask, default router IP address, optional VLAN identifier, and port number of the AP manager interface
	
	Assigns the IP address of the DHCP server that will assign addresses to the APs

	NOTE: The AP manager interface is used for Layer 3 communication between the controller and the lightweight APs. It must have a unique IP address and is usually configured as the same VLAN or IP subnet as the management interface, but is not required to be.
	NOTE: If the AP manager interface is on the same subnet as the management interface, the same DHCP server is used for the AP manager interface and the management interface.

`Virtual Gateway IP Address: `**`1.1.1.1`**	Assigns an IP address to the controller's virtual interface.
	NOTE: Because the virtual gateway will not be routed, use a fictitious, unassigned IP address, such as 1.1.1.1. All controllers within a mobility group must be configured with the same virtual interface IP address.
`Mobility/RF Group Name: `**`bcmsn`**	Assigns the name of the mobility group/RF group to which this controller belongs
`Network Name (SSID): `**`bcmsn`**	Assigns the service set identifier (SSID)
`Allow Static IP Addresses [YES][no]:`	Enter **yes** (or press the ⏎Enter key) if you want to allow clients to assign their own IP addresses. Enter **no** if you want to force clients to use IP addresses from a DHCP server.

`Configure a RADIUS Server now? [YES][no]: `**`no`**
`Warning! The default WLAN security policy`
`requires a RADIUS server.`
`Please see documentation for more details.`
	Enter **yes** if you want to configure a RADIUS server now. Enter **no** if you do not want to do so at this time.

`Enter Country Code (enter 'help' for a list of` `countries) [US]: `**`US`**	Assigns the country code

`Enable 802.11b Network [YES][no]:`
`Enable 802.11a Network [YES][no]:`
`Enable 802.11g Network [YES][no]:`
	Enter **yes** (or press the ⏎Enter key) to enable each of the 802.11b, a, or g lightweight AP networks. Enter **no** to disable.

`Enable Auto-RF [YES][no]:`	Enter **yes** to enable or **no** to disable the Radio Resource Management (RRM) auto-RF feature
	NOTE: The auto-RF feature enables the controller to automatically form an RF group with other controllers.

`Configuration saved!`
 	
`Resetting system with new configuration...`
	Controller saves configuration and then automatically reboots the controller

	NOTE: This is the end of the wizard. From here on are examples of commands used to continue with the configuration and verification of the controller.
`<Output Cut>`	

`(Cisco Controller)`
 	
`Enter User Name (or 'Recover-Config' this one-time only to reset configuration to factory defaults)`
 	
`User: cisco`
 	
`Password:********`
	Enter your username and password to log in to the controller

`(Cisco Controller) >config prompt` `WLAN_Controller`	Changes the controller prompt to WLAN_Controller. The length of this prompt is 31 alphanumeric characters.

(WLAN_Controller) >`config network telnet enable`	Enables Telnet access to the WLAN controller. By default, controllers block Telnet connections.
(WLAN_Controller) >`config network webmode enable`	Enables HTTP access to the WLAN controller
	NOTE: HTTPS access is enabled by default; unsecured HTTP is not
(WLAN_Controller) >`config serial timeout 3`	Sets automatic logout of the CLI to 3 minutes
	NOTE: The default timeout for the CLI is 0 minutes. The range of the **config serial timeout** command is 0 to 160, measured in minutes, where 0 represents never logging out.
(WLAN_Controller) >`save config` Are you sure you want to save? (y/n) `y` Configuration Saved!	Saves the configuration
(WLAN_Controller) >`show interface summary`	Displays current interface configuration
(WLAN_Controller) >`show run-config`	Displays current configuration
(WLAN_Controller) >`show ap summary`	Displays a summary of all Cisco 1000 series lightweight APs attached to the controller
(WLAN_Controller) >`show wlan summary`	Displays a summary of the WLANs
(WLAN_Controller) > `show port summary`	Displays the status of the controller's distribution system ports

After configuration is complete, you can open up a web browser and connect to the device. Figure 6-1 shows the GUI login screen. Using the preceding configuration as a guide, you would connect to 172.16.1.100. If you are connecting to an unconfigured controller, you would use the address 192.168.1.1.

Figure 6-1   GUI Login Screen

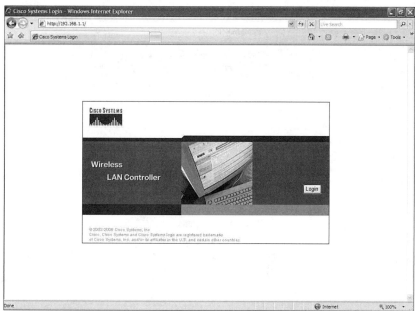

Figure 6-2 shows the Login screen once the Login button has been pressed.

Figure 6-2   GUI Login Screen After the Login Button Has Been Pressed

Figure 6-3 shows the main page after a successful login.

Figure 6-3   Main Page

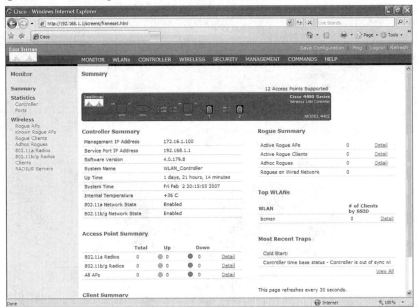

Configuration Example: 4402 WLAN Controller Using the Web Interface

NOTE: The Cisco 4400 series WLAN controller supports the initial configuration via a web browser through the service port. The default address of the unconfigured controller is 192.168.1.1. The default username and password are both admin.

Cisco recommends using Internet Explorer 6.0 with Service Pack 1 (SP1) or later with full switch web interface functionality.

There are known issues with Opera, Mozilla, and Netscape.

Refer back to Figure 6-1 and Figure 6-2 for the GUI Login screen. If you use the default address of 192.168.1.1 and the default username/password combination of admin/admin, the GUI Configuration Wizard will appear. Figure 6-4 shows the first screen of the GUI Configuration Wizard.

Figure 6-4   First Screen of the GUI Configuration Wizard

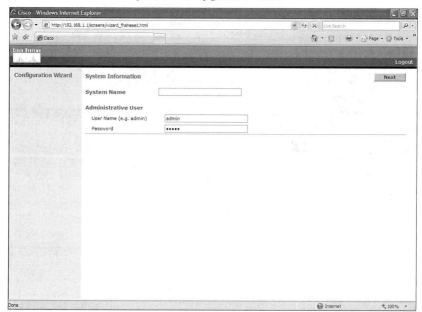

Figure 6-5 shows the second screen of the GUI Configuration Wizard. This is where you configure the IP address and netmask of the service interface and enable DHCP, if desired.

Figure 6-5   Service Interface Configuration of the GUI Configuration Wizard

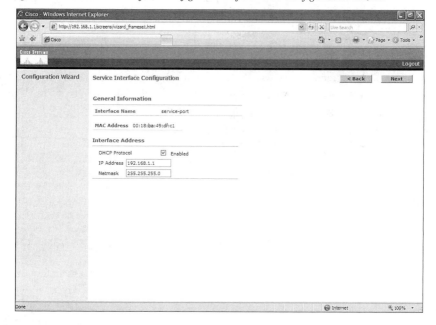

Figure 6-6 shows the third screen of the GUI Configuration Wizard. This is where you assign the IP address, netmask, default router IP address, optional VLAN identifier of the management interface, and port number of the management interface. You also configure the IP address of the DHCP server that will assign addresses to the APs. Note that if you leave the VLAN identifier as zero, it means the interface is untagged.

Figure 6-6   Management Interface Configuration Screen of the GUI Configuration Wizard

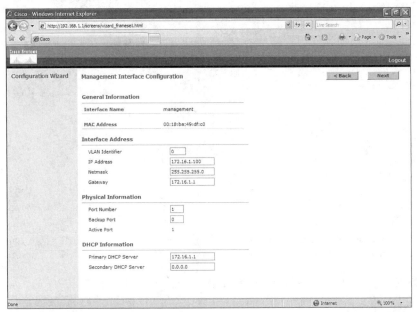

Figure 6-7 shows the fourth screen of the GUI Configuration Wizard. This is where you set the AP transport layer mode along with the RF mobility domain name and the country code. Note that the screen scrolls down to list more country codes

Figure 6-8 shows the fifth screen of the GUI Configuration Wizard. This is where you assign the IP address, netmask, default router IP address, optional VLAN identifier, and port number of the AP Manager Interface. You also assign the address of the DHCP server. If the AP manager interface is on the same subnet as the management interface, the same DHCP server is used for the AP manager interface and the management interface.

Figure 6-7   Miscellaneous Configuration of the GUI Configuration Wizard

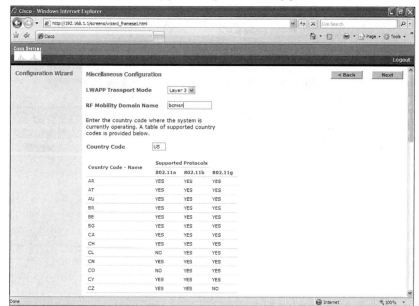

Figure 6-8   AP Manager Interface Configuration Screen of the GUI Configuration Wizard

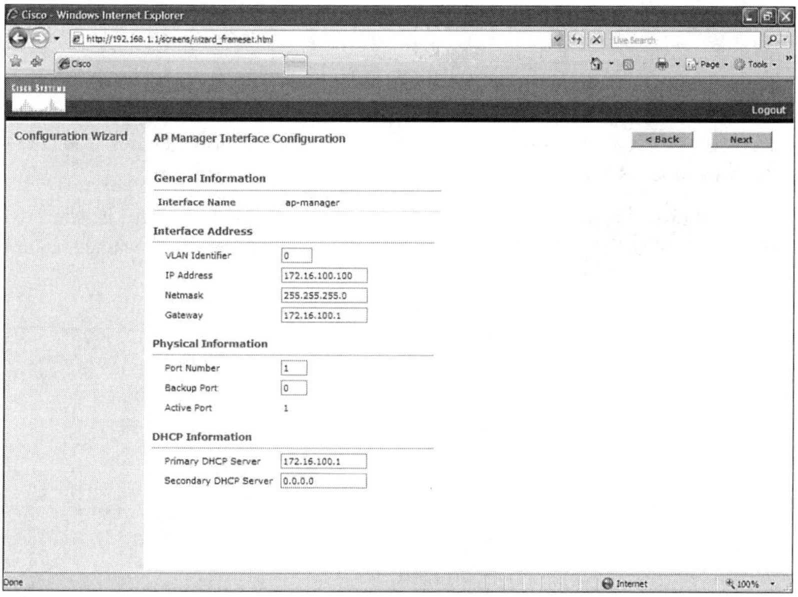

Figure 6-9 shows the sixth screen of the GUI Configuration Wizard. This is where you assign the IP address of the virtual interface. Because the virtual gateway will not be routed, use a fictitious, unassigned IP address, such as 1.1.1.1. All controllers within a mobility group must be configured with the same virtual interface IP address.

Figure 6-9   Virtual Interface Configuration Screen of the GUI Configuration Wizard

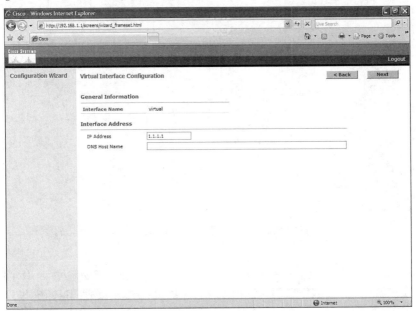

Figure 6-10 shows the seventh screen of the GUI Configuration Wizard. This is where you configure the WLAN SSID, along with general policies and security policies. You also set the 802.1x parameters here.

Figure 6-11 shows the eighth screen of the GUI Configuration Wizard. This is where you configure your RADIUS server. In this example, no RADIUS server was wanted, so this screen was left to the default settings.

Figure 6-10   WLAN Policy Configuration Screen of the GUI Configuration Wizard

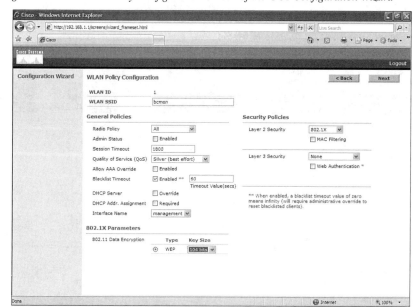

Figure 6-11   RADIUS Server Configuration Screen of the GUI Configuration Wizard

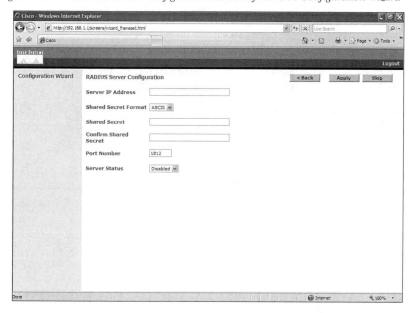

Figure 6-12 shows the ninth screen of the GUI Configuration Wizard. This is where you enable the network status of your wireless technologies—802.11a/b/g and Auto-RF.

Figure 6-12   802.11 Configuration Screen of the GUI Configuration Wizard

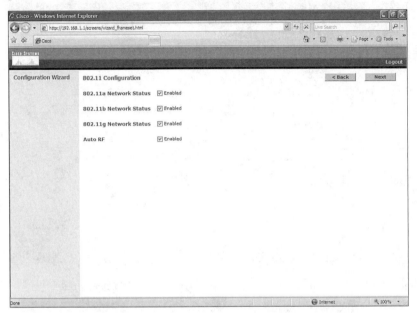

Figure 6-13 shows the tenth screen of the GUI Configuration Wizard. At this point, the configuration is complete. The pop-up will appear after you click the Save and Reboot button. The configuration will save, and the controller will then restart.

After the system has been rebooted, HTTP will no longer work. You must use HTTPS. Figure 6-14 shows the Login screen in HTTPS.

Figure 6-13   Configuration Wizard Complete Screen of the GUI Configuration Wizard

Figure 6-14   Login Screen in HTTPS

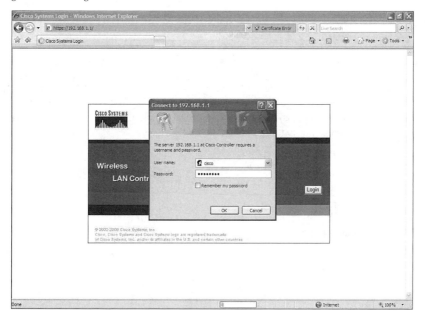

To enable HTTP access, choose the Management tab on the top of the page, and then select HTTP on the left side of the screen, as illustrated in Figure 6-15. Choose the Enabled option for HTTP Access. Note that in Figure 6-15, HTTP has not been enabled yet, but HTTPS has.

Figure 6-15   Enabling HTTP Access

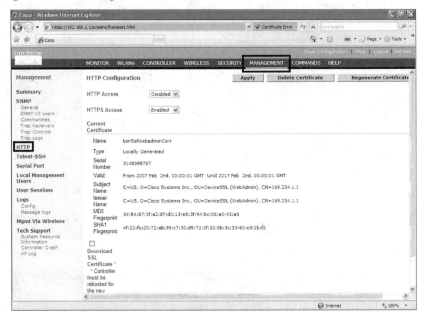

To change the controller name, choose the Management tab on the top of the page, and then select SNMP on the left side of the screen, as illustrated in Figure 6-16. Here you can change the controller name, add a description of the location of the controller, and add the contact information of the controller administrator.

Figure 6-16   Changing Controller Name

Figure 6-17 shows a summary of the menu bar in the GUI of the WLC.

Figure 6-17   WLC Web Menu Bar

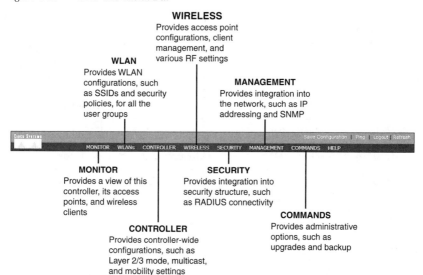

Configuration Example: Configuring a 3560 Switch to Support WLANs and APs

Figure 6-18 shows the network topology for the configuration that follows, which shows how to configure a 3560 switch to support WLANs and APs.

Figure 6-18   Topology for WLAN/AP Support Configuration on a 3560 Switch

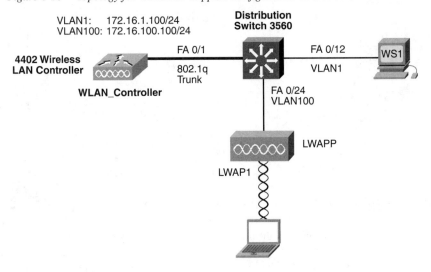

`Switch>`**`enable`**	Moves to privileged mode
`Switch#`**`configure terminal`**	Moves to global configuration mode
`Switch(config)#`**`hostname 3560`**	Sets host name of switch
`3560(config)#`**`vlan 1`**	Enters VLAN-configuration mode
`3560(config-vlan)#`**`name Management`**	Assigns a name to VLAN 1
`3560(config-vlan)#`**`exit`**	Returns to global config mode
`3560(config)#`**`vlan 100`**	Creates VLAN 100 and enters VLAN-config mode
`3560(config-vlan)#`**`name Wireless`**	Assigns a name to VLAN 100
`3560(config-vlan)#`**`exit`**	Returns to global config mode
`3560(config)#`**`interface vlan 1`**	Moves to interface configuration mode
`3560(config-if)#`**`ip address 172.16.1.1 255.255.255.0`**	Assigns IP address and netmask

3560(config-if)#no shutdown	Enables the interface
3560(config-if)#interface vlan 100	Moves to interface config mode
3560(config-if)#ip address 172.16.100.1 255.255.255.0	Assigns IP address and netmask
3560(config-if)#no shutdown	Enables the interface
3560(config-if)#exit	Returns to global config mode
3560(config)#ip dhcp pool wireless	Creates a DHCP pool called wireless and enters DHCP configuration mode
3560(config-dhcp)#network 172.16.100.0 255.255.255.0	Defines the range of addresses to be leased
3560(config-dhcp)#default router 172.16.100.1	Defines the address of the default router for the client
3560(config-dhcp)#exit	Returns to global config mode
3560(config)#interface fastethernet 0/1	Moves to interface config mode
3560(config-if)#description link to WLAN_Controller	Creates locally significant description
3560(config-if)#switchport mode trunk	Makes this interface a trunk port
3560(config-if)#interface fastethernet 0/24	Moves to interface config mode for interface fastethernet 0/24
3560(config-if)#description link to Access Point	Creates locally significant description
3560(config-if)#switchport mode access	Makes this interface an access port
3560(config-if)#switchport access vlan 100	Assigns this interface to VLAN 100
3560(config-if)#spanning-tree portfast	Enables PortFast on this interface
3560(config-if)#exit	Returns to global config mode
3560(config)#exit	Returns to privileged mode
3560#copy running-config startup-config	Saves the configuration to NVRAM

Configuration Example: Configuring a Wireless Client

Refer back to Figure 6-18, which shows the network topology for the following configuration on how to configure a Cisco Aironet wireless client adapter:

Step 1. Install a Cisco Aironet Wireless Adapter into an open slot on your laptop.

Step 2. Load the Cisco Aironet Desktop Utility software onto your laptop.

Step 3. If necessary, reboot your machine, and then run the Aironet Desktop Utility program.

Step 4. Open the Profiles Management tab and click New (see Figure 6-19).

Figure 6-19   Profile Management Screen

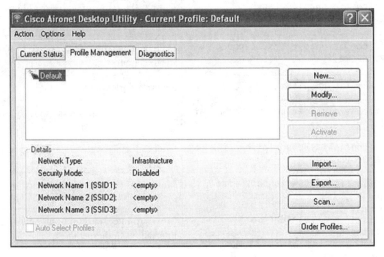

Step 5. Enter your profile name, client name, and SSID (see Figure 6-20).

Step 6. Open the Security tab and choose None (see Figure 6-21).

Figure 6-20   SSID Configuration

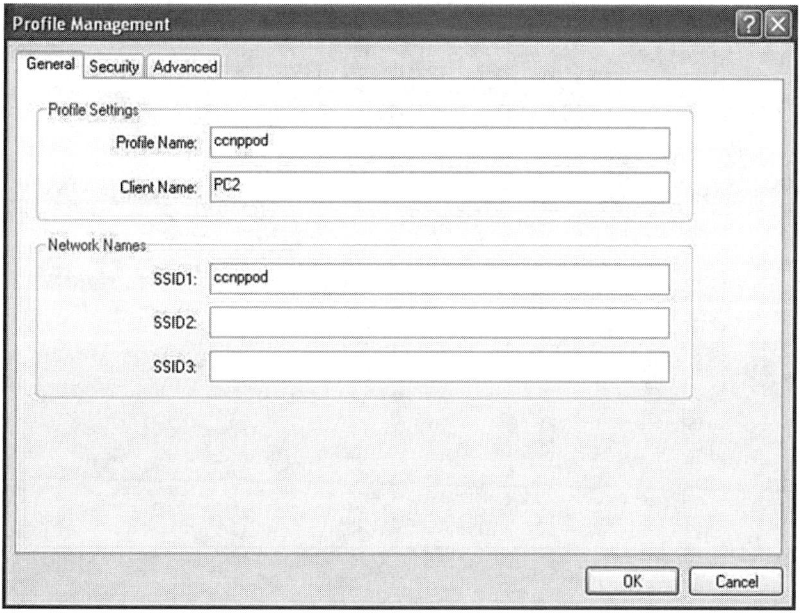

Figure 6-21   Security Options

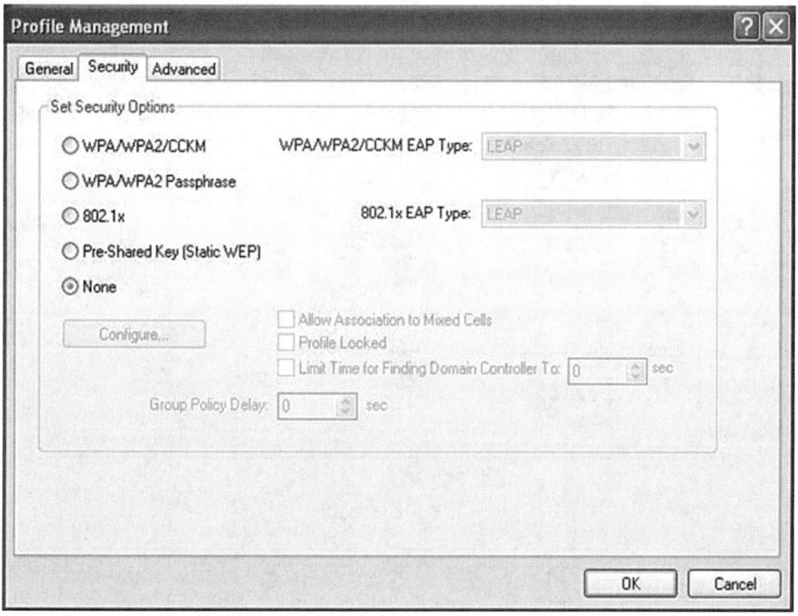

Step 7. Open the Advanced tab. Uncheck the 5GHz 54 Mbps, because you are not using 802.11a. Then click OK (see Figure 6-22).

Figure 6-22   Advanced Options

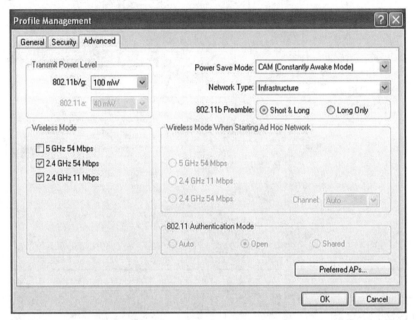

Step 8. After clicking OK, you are returned to the Profile Management screen. In addition to the default profile, there is a new profile called ccnppod. Select the ccnppod profile and click the Activate button. After clicking the Activate button, the screen will look like Figure 6-23.

Figure 6-23   ccnppod Profile Activated

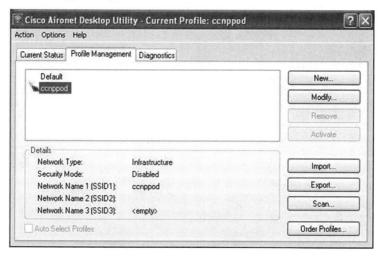

Step 9.   Click the Current Status tab, and your screen should look similar to Figure 6-24.

Figure 6-24   Current Status of ccnppod Profile

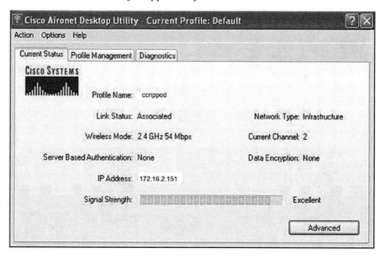

CHAPTER 7

Minimizing Service Loss and Data Theft

This chapter provides information and commands concerning the following topics:

- Configuring static MAC addresses
- Switch port security
- Verifying switch port security
- Sticky MAC addresses
- Mitigating VLAN hopping: best practices
- Configuring private VLANs (PVLAN)
- Verifying PVLANs
- Configuring protected ports
- VLAN access maps
- Verifying VLAN access maps
- DHCP snooping
- Verifying DHCP snooping
- Dynamic ARP Inspection (DAI)
- Verifying DAI
- 802.1x port-based authentication
- Cisco Discovery Protocol (CDP) security issues
- Configuring the Secure Shell (SSH) protocol
- vty access control lists (ACL)
- Restricting web interface sessions with ACLs
- Disabling unneeded services
- Securing end-device access ports

Configuring Static MAC Addresses

You can define the forwarding behavior of a switch port by adding a static MAC address to your configuration. This MAC address can be either a unicast or a multicast address, and the entry does not age and is retained when the switch restarts.

`Switch(config)#macaddress-table`
`static cf23.1943.9a4b vlan 1`
`interface fastethernet 0/3`
	Destination MAC address cf23.1943.9a4b is added to the MAC address table. Packets with this address are forwarded out interface fastethernet 0/3.

	NOTE: Beginning with Cisco IOS Software Release 12.1(11)EA1, the **mac address-table static** command (no hyphen) replaces **the mac-address-table** command (with the hyphen). The **mac-address-table static** command (with the hyphen) will become obsolete in a future release.

`Switch(config)#mac address-table`
`static 1234.5678.90ab vlan 4`
`interface gigabitethernet 0/1`
	Destination MAC address 1234.5678.90ab is added to the MAC address table. Packets with this address and forwarded out interface gigabitethernet 0/1.

Switch Port Security

`Switch(config)#interface` `fastethernet 0/1`	Moves to interface configuration mode
`Switch(config-if)#switchport port-` `security`	Enables port security on the interface
`Switch(config-if)#switchport port-` `security maximum 4`	Sets a maximum limit of 4 MAC addresses that will be allowed on this port
	NOTE: The maximum number of secure MAC addresses that you can configure on a switch is set by the maximum number of available MAC addresses allowed in the system.

`Switch(config-if)#switchport port-`
`security mac-address`
`1234.5678.90ab`
	Sets a specific secure MAC address 1234.5678.90ab. You can add additional secure MAC addresses up to the maximum value configured.

`Switch(config-if)#switchport port-` `security violation shutdown`	Configures port security to shut down the interface if a security violation occurs

	NOTE: In shutdown mode, the port is errdisabled, a log entry is made, and manual intervention or errdisable recovery must be used to reenable the interface.
`Switch(config-if)#`**`switchport port-`****`security violation restrict`**	Configures port security to restrict mode if a security violation occurs
	NOTE: In restrict mode, frames from a nonallowed address are dropped, and a log entry is made. The interface remains operational.
`Switch(config-if)#`**`switchport port-`****`security violation protect`**	Configures port security to protect mode if a security violation occurs
	NOTE: In protect mode, frames from a nonallowed address are dropped, but no log entry is made. The interface remains operational.

Verifying Switch Port Security

`Switch#`**`show port-security`**	Displays security information for all interfaces
`Switch#`**`show port-security`****`interface fastethernet 0/5`**	Displays security information for interface fastethernet 0/5
`Switch#`**`show port-security address`**	Displays MAC address table security Information
`Switch#`**`show mac address-table`**	Displays the MAC address table
`Switch#`**`clear mac address-table`****`dynamic`**	Deletes all dynamic MAC addresses
`Switch#`**`clear mac address-table`****`dynamic address aaaa.bbbb.cccc`**	Deletes the specified dynamic MAC address
`Switch#`**`clear mac address-table`****`dynamic interface fastethernet 0/5`**	Deletes all dynamic MAC addresses on interface fastethernet 0/5
`Switch#`**`clear mac address-table`****`dynamic vlan 10`**	Deletes all dynamic MAC addresses on VLAN 10

`Switch#clear mac address-table notification`	Clears MAC notification global counters
	NOTE: Beginning with Cisco IOS Software Release 12.1(11)EA1, the **clear mac address-table** command (no hyphen) replaces the **clear mac-address-table** command (with the hyphen). The **clear mac-address-table static** command (with the hyphen) will become obsolete in a future release.

Sticky MAC Addresses

Sticky MAC addresses are a feature of port security. Sticky MAC addresses limit switch port access to a specific MAC address that can be dynamically learned, as opposed to a network administrator manually associating a MAC addresses with a specific switch port. These addresses are stored in the running configuration file. If this file is saved, the sticky MAC addresses will not have to be relearned when the switch is rebooted, providing a high level of switch port security.

`Switch(config)#interface fastethernet 0/5`	Moves to interface config mode
`Switch(config-if)#switchport port-security mac-address sticky`	Converts all dynamic port security–learned MAC addresses to sticky secure MAC addresses
`Switch(config-if)#switchport port-security mac-address sticky vlan 10 voice`	Converts all dynamic port security–learned MAC addresses to sticky secure MAC addresses on voice VLAN 10
	NOTE: The **voice** keyword is available only if a voice VLAN is first configured on a port and if that port is not the access VLAN.

Mitigating VLAN Hopping: Best Practices

Configure all unused ports as access ports so that trunking cannot be negotiated across those links.

Place all unused ports in the shutdown state and associate with a VLAN designed only for unused ports, carrying no user data traffic.

When establishing a trunk link, purposefully configure the following:

- The native VLAN to be different from any data VLANs
- Trunking as **on**, rather than **negotiated**
- The specific VLAN range to be carried on the trunk

Configuring Private VLANs

A problem can potentially exist when an Internet service provider (ISP) has many devices from different customers on a single demilitarized zone (DMZ) segment or VLAN—these devices are not isolated from each other. Some switches can implement private VLANs (PVLAN), which will keep some switch ports shared and some isolated, even though all ports are in the same VLAN. This isolation eliminates the need for a separate VLAN and IP subnet per customer.

> **NOTE:** Private VLANs are implemented on Catalyst 6500/4500/3750/3560 switches.

`Switch(config)#`**`vtp mode transparent`**	Sets VLAN Trunking Protocol (VTP) mode to transparent
`Switch(config)#`**`vlan 20`**	Creates VLAN 20 and moves to VLAN-configuration mode
`Switch(config-vlan)#`**`private-vlan primary`**	Creates a private, primary VLAN
`Switch(config-vlan)#`**`vlan 101`**	Creates VLAN 101 and moves to VLAN-config mode
`Switch(config-vlan)#`**`private-vlan isolated`**	Creates a private, isolated VLAN for VLAN 101
	NOTE: An isolated VLAN can only communicate with promiscuous ports.
`Switch(config-vlan)#`**`exit`**	Returns to global configuration mode
`Switch(config)#`**`vlan 102`**	Creates VLAN 102 and moves to VLAN-config mode

`Switch(config-vlan)#`**`private-vlan`** **`community`**	Creates a private, community VLAN for VLAN 102
	NOTE: A community VLAN can communicate with all promiscuous ports and with other ports in the same community.
`Switch(config-vlan)#`**`exit`**	Returns to global config mode
`Switch(config)#`**`vlan 103`**	Creates VLAN 103 and moves to VLAN-config mode
`Switch(config-vlan)#`**`private-vlan`** **`community`**	Creates a private, community VLAN for VLAN 103
`Switch(config-vlan)#`**`vlan 20`**	Returns to VLAN-config mode for VLAN 20
`Switch(config-vlan)#`**`private-vlan`** **`association 101-103`**	Associates secondary VLANs 101–103 with primary VLAN 20
	NOTE: Only one isolated VLAN can be mapped to a primary VLAN, but more than one community VLAN can be mapped to a primary VLAN.
`Switch(config)#`**`interface`** **`fastethernet 0/20`**	Moves to interface config mode
`Switch(config-if)#`**`switchport`** **`mode private-vlan host`**	Configures the port as a private VLAN host port
`Switch(config-if)#`**`switchport`** **`private-vlan host-association 20`** **`101`**	Associates the port with primary private VLAN 20 and secondary private VLAN 101

Verifying PVLANs

`Switch#`**`show vlan private-vlan`** **`type`**	Verifies private VLAN configuration
`Switch#`**`show interface`** **`fastethernet 0/20 switchport`**	Verifies all configuration on fastethernet 0/20, including private VLAN associations

Configuring Protected Ports

NOTE: Although 2960/3560 switches do not support private VLANs, they do support protected ports, which has a similar functionality to PVLANs on a per-switch basis.

`Switch(config)#interface fastethernet 0/17`	Moves to interface config mode
`Switch(config-if)#switchport protected`	Configures interface to be a protected port
	NOTE: The use of protected ports ensures that there is no exchange of unicast, broadcast, or multicast traffic between these ports on the switch. Only control traffic will be forwarded. All data traffic passing between protected ports must be forwarded through a Layer 3 device.
`Switch(config-if)#exit`	Returns to global config mode

VLAN Access Maps

VLAN access maps are the only way to control filtering within a VLAN. VLAN access maps have no direction—if you want to filter traffic in a specific direction, you need to include an access control list (ACL) with specific source or destination addresses. VLAN access maps do not work on the 2960 platform, but they do work on the 3560 and the 6500 platforms.

`3560Switch(config)#ip access-list extended test1`	Creates a named extended ACL called test1
`3560Switch(config-ext-nacl)#permit tcp any any`	The first line of an extended ACL will permit any TCP packet from any source to travel to any destination address. Because there is no other line in this ACL, the implicit **deny** statement that is part of all ACLS will deny any other packet.
`3560Switch(config-ext-nacl)#exit`	Exits named ACL configuration mode and returns to global config mode

`3560Switch(config)#`**`vlan access-`** **`map drop_TCP`**	Creates a VLAN access map named drop_TCP and moves into VLAN access map configuration mode. If no sequence number is given at the end of the command, a default number of 10 is assigned.
`3560Switch(config-access-` `map)#`**`match ip address test1`**	Defines what needs to occur for this action to continue. In this case, packets filtered out by the named ACL test1 will be acted upon.
	NOTE: You can match ACLs based on the following: IP ACL number: 1–199 and 1300–2699 IP ACL name IPX ACL number: 800–999 IPX ACL name MAC address ACL name
`3560Switch(config-access-` `map)#`**`action drop`**	Any packet that is filtered out by the ACL test1 will be dropped
	NOTE: You can configure the following actions: Drop Forward Redirect (works only on a Catalyst 6500)
`3560Switch(config-access-` `drop)#`**`exit`**	Exits access map configuration mode and returns to global config mode
`3560Switch(config)#`**`vlan filter`** **`drop_TCP vlan-list 20-30`**	Applies the VLAN map named drop_TCP to VLANs 20–30
	NOTE: The **vlan-list** argument can refer to a single VLAN (26), a consecutive list (20–30), or a string of VLAN IDs (12, 22, 32). Spaces around the comma and hyphen are optional.

Verifying VLAN Access Maps

Switch#**show vlan access-map**	Displays all VLAN access maps
Switch#**show vlan access-map drop_TCP**	Displays the VLAN access map named drop_TCP
Switch#**show vlan filter**	Displays what filters are applies to all VLANs
Switch#**show vlan filter access-map drop_TCP**	Displays the filter for the specific VLAN access map named drop_TCP

Configuration Example: VLAN Access Maps

Figure 7-1 shows the network topology for the configuration that follows, which shows how to configure VLAN access maps using the commands covered in this chapter.

Figure 7-1   Network Topology for VLAN Access Map Configuration

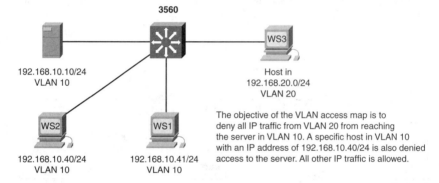

The objective of the VLAN access map is to deny all IP traffic from VLAN 20 from reaching the server in VLAN 10. A specific host in VLAN 10 with an IP address of 192.168.10.40/24 is also denied access to the server. All other IP traffic is allowed. A 3560 switch is used for this example.

`3560(config)#ip access-list extended DENY_SERVER_ACL`	Creates a named ACL called DENY_SERVER_ACL and moves to named ACL configuration mode
`3560(config-ext-nacl)#permit ip 192.168.20.0 0.0.0.255 host 192.168.10.10`	This line filters out all IP packets from a source address of 192.168.20.x destined for the server at 192.168.10.10
`3560(config-ext-nacl)#permit ip host 192.168.10.40 host 192.168.10.10`	This line filters out all IP packets from a source address of 192.168.10.40 destined for the server at 192.168.10.10
`3560(config-ext-nacl)#exit`	Returns to global config mode
`3560(config)#vlan access-map DENY_SERVER_MAP 10`	Creates a VLAN access map called DENY_SERVER_MAP and moves into VLAN access map config mode. If no sequence number is given at the end of the command, a default number of 10 is assigned.
`3560(config-access-map)#match ip address DENY_SERVER_ACL`	Defines what needs to occur for this action to continue. In this case, packets filtered out by the named ACL DENY_SERVER_ACL will be acted upon.
`3560(config-access-map)#action drop`	Any packet filtered out by the ACL will be dropped
`3560(config-access-map)#exit`	Returns to global config mode
`3560(config)#vlan access-map DENY_SERVER_MAP 20`	Creates line 20 of the VLAN access map called DENY_SERVER_MAP and moves into VLAN access map config mode
`3560(config-access-map)#action forward`	Any packet not filtered out by the ACL in line 10 will be forwarded
`3560(config-access-map)#exit`	Returns to global config mode
`3560(config)#vlan filter DENY_SERVER_MAP vlan-list 10`	Applies the VLAN map to VLAN 10

DHCP Snooping

DHCP snooping is a DHCP security feature that provides network security by filtering untrusted DHCP messages and by building and maintaining a DHCP snooping binding database, which is also referred to as a DHCP snooping binding table.

`Switch(config)#ip dhcp snooping`	Enables DHCP snooping globally
	NOTE: If you enable DHCP snooping on a switch, the following DHCP relay agent commands are not available until snooping is disabled: `Switch(config)#ip dhcp relay information check` `Switch(config)#ip dhcp relay information policy {drop ¦ keep ¦ replace}Switch(config)#ip dhcp relay information trust-all` `Switch(config-if)#ip dhcp relay information trusted` If you enter these commands with DHCP snooping enabled, the switch returns an error message.
`Switch(config)#ip dhcp snooping vlan 20`	Enables DHCP snooping on VLAN 20
`Switch(config)#ip dhcp snooping vlan 10-35`	Enables DHCP snooping on VLANs 10–35
`Switch(config)#ip dhcp snooping vlan 20 30`	Enables DHCP snooping on VLANs 20–30
`Switch(config)#ip dhcp snooping 10,12,14`	Enables DHCP snooping on VLANs 10, 12, and 14
`Switch(config)#ip dhcp snooping information option`	Enables DHCP option 82 insertion

	NOTE: DHCP address allocation is usually based on an IP address—either the gateway IP address or the incoming interface IP address. In some networks, you might need additional information to determine which IP address to allocate. By using the "relay agent information option"—option 82—the Cisco IOS relay agent can include additional information about itself when forwarding DHCP packets to a DHCP server. The relay agent will add the circuit identifier suboption and the remote ID suboption to the relay information option and forward this all to the DHCP server.
`Switch(config)#`**`interface fasthethernet 0/1`**	Moves to interface config mode
`Switch(config-if)#`**`ip dhcp snooping trust`**	Configures the interface as trusted
	NOTE: There must be at least one trusted interface when working with DHCP snooping. It is usually the port connected to the DHCP server or to uplink ports. By default, all ports are untrusted.
`Switch(config-if)#`**`ip dhcp snooping limit rate 75`**	Configures the number of DHCP packets per second that an interface can receive
	NOTE: The range of packets that can be received per second is 1 to 4,294,967,294. The default is no rate configured.
	TIP: Cisco recommends an untrusted rate limit of no more than 100 packets per second.
`Switch(config-if)#`**`ip dhcp snooping verify mac-address`**	Configures the switch to verify that the source MAC address in a DHCP packet that is received on an untrusted port matches the client hardware address in the packet

Verifying DHCP Snooping

Switch#show ip dhcp snooping	Displays the DHCP snooping configuration for a switch
Switch#show ip dhcp snooping binding	Displays only the dynamically configured bindings in the DHCP snooping binding database
Switch#show running-config	Displays the status of the insertion and removal of the DHCP option 82 field on all interfaces

Dynamic ARP Inspection

Dynamic ARP Inspection (DAI) determines the validity of an ARP packet. This feature prevents attacks on the switch by not relaying invalid ARP requests and responses to other ports in the same VLAN. DAI does not work on the 2960.

NOTE: To use this feature, you must have the enhanced multilayer image (EMI) installed on your 3560 switch.

3560Switch(config)#ip arp inspection vlan 10	Enables DAI on VLAN 10
3560Switch(config)#ip arp inspection vlan 10,20	Enables DAI on VLANs 10 and 20
3560Switch(config)#ip arp inspection vlan 10-20	Enables DAI on VLANs 10 to 20 inclusive
3560Switch(config)#ip arp inspection validate src-mac	Configures DAI to drop ARP packets when the source MAC address in the body of the ARP packet does not match the source MAC address specified in the Ethernet header. This check is performed on both APR requests and responses.
3560Switch(config)#ip arp inspection validate dst-mac	Configures DAI to drop ARP packets when the destination MAC address in the body of the ARP packet does not match the destination MAC address specified in the Ethernet header. This check is performed on both APR requests and responses.

`3560Switch(config)#ip arp inspection validate ip`	Configures DAI to drop ARP packets that have invalid and unexpected IP addresses in the ARP body, such as 0.0.0.0, 255.255.255.255, or all IP multicast addresses. Sender IP addresses are checked in all ARP requests and responses, and target IP addresses are checked only in ARP responses.
`Switch(config)#interface fastethernet 0/24`	Moves to interface config mode
`Switch(config-if)#ip arp inspection trust`	Configures the connection between switches as trusted
	NOTE: By default, all interfaces are untrusted.

Verifying DAI

`Switch#show ip arp inspection interfaces`	Verifies the dynamic ARP configuration
`Switch#show ip arp inspection vlan 10`	Verifies the dynamic ARP configuration for VLAN 10
`Switch#show ip arp inspection statistics vlan 10`	Displays the dynamic ARP inspection statistics for VLAN 10

802.1x Port-Based Authentication

The IEEE 802.1x standard defines an access control and authentication protocol that prevents unauthorized hosts from connecting to a LAN through publicly accessible ports unless they are properly authenticated. The authentication server authenticates each host connected to a switch port before making available any services offered by the switch or the LAN.

`Switch(config)#aaa new-model`	Enables authentication, authorization, and accounting (AAA)
`Switch(config)#aaa authentication dot1x default group radius`	Creates an 802.1x port-based authentication method list. This method specifies using a RADIUS server for authentication.

	NOTE: A method list describes the sequence and authentication methods to be queried to authenticate a user. The software uses the first method listed to authenticate users; if that method fails to respond, the software selects the next authentication method in the method list. This process continues until there is successful communication with a listed authentication method or until all defined methods are exhausted. If authentication fails at any point in this cycle, the authentication process stops, and no other authentication methods are attempted.
	NOTE: To create a default list that is used when a named list is not specified, use the **default** keyword followed by methods that are to be used in default situations.
	NOTE: When using the **aaa authentication dot1x** command, you must use at least one of the following keywords: **group radius**—Use a list of RADIUS servers for authentication. **none**—Use no authentication. The client is automatically authenticated without the switch using information supplied by the client. This method should only be used as a second method. If the first method of **group radius** is not successful, the switch will use the second method for authentication until a method is successful. In this case, no authentication would be used.
`Switch(config)#`**`dot1x system-auth-control`**	Globally enables 802.1x port-based authentication
`Switch(config)#`**`interface fastethernet 0/1`**	Moves to interface config mode
`Switch(config-if)#`**`dot1x port-control auto`**	Enables 802.1x authentication on this interface

	NOTE: The **auto** keyword allows the port to begin in the unauthorized state. This will allow only Extensible Authentication Protocol over LAN (EAPOL) frames to be sent and received through the port. Other keywords available here are these: **force-authorized**—Disables 802.1x authentication and causes the port to transition to the authorized state without any authentication exchange required. This is the default setting. **force-unauthorized**—Causes the port to remain in the unauthorized state, ignoring all attempts by the client to authenticate. The switch cannot provide authentication services to the client through the interface.
Switch#**show dot1x**	Verifies your 802.1x entries

Cisco Discovery Protocol Security Issues

Although Cisco Discovery Protocol (CDP) is necessary for some management applications, CDP should still be disabled in some instances.

Disable CDP globally under these scenarios:

- CDP is not required at all.
- The device is located in an insecure environment.

Use the command **no cdp run** to disable CDP globally:

```
Switch(config)#no cdp run
```

Disable CDP on any interface under these scenarios:

- Management is not being performed.
- The interface is a nontrunk interface.
- The interface is connected to a nontrusted network.

Use the interface configuration command **no cdp enable** to disable CDP on a specific interface:

```
Switch(config)#interface fastethernet 0/12
Switch(config-if)#no cdp enable
```

Configuring the Secure Shell Protocol

CAUTION: Secure Shell (SSH) Version 1 implementations have known security issues. It is recommended to use SSH Version 2 whenever possible.

NOTE: To work, SSH requires a local username database, a local IP domain, and an RSA key to be generated.

The Cisco implementation of SSH requires Cisco IOS Software to support Rivest, Shamit, Adleman (RSA) authentication and minimum Data Encryption Standard (DES) encryption—a cryptographic software image.

`Switch(config)#username Roland password tower`	Creates a locally significant username/password combination. These are the credentials needed to be entered when connecting to the switch with SSH client software.
`Switch(config)#ip domain-name test.lab`	Creates a host domain for the switch
`Switch(config)#crypto key generate rsa`	Enables the SSH server for local and remote authentication on the switch and generates an RSA key pair

vty ACLs

`Switch(config)#access-list 10 permit host 192.168.1.15`	Creates a standard ACL that filters out traffic from source address 192.168.1.15
`Switch(config)#line vty 0 15`	Moves to vty line mode. All commands in this mode will apply to vty lines 0–15 inclusive.
`Switch(config-line)#access-class 10 in`	Restricts incoming vty connections to addresses filtered by ACL 10
	NOTE: The actual number of vty lines depends on the platform and the version of Cisco IOS Software being run.

Restricting Web Interface Sessions with ACLs

Switch(config)#**access-list 10 permit host 192.168.1.15**	Creates a standard ACL that filters out traffic from source address 192.168.1.15
Switch(config)#**ip http server**	Enables the HTTP server on the switch
Switch(config)#**ip http access-class 10**	Applies ACL 10 to the HTTP server

Disabling Unneeded Services

TIP: Cisco devices implement various TCP and User Datagram Protocol (UDP) servers to help facilitate management and integration of devices. If these servers are not needed, consider disabling them to reduce security vulnerabilities.

Switch(config)#**no service tcp-small-servers**	Disables minor TCP services—echo, discard, chargen, and daytime—available from hosts on the network
Switch(config)#**no service udp-small-servers**	Disables minor UDP services—echo, discard, and chargen—available from hosts on the network
Switch(config)#**no ip finger**	Disables the finger service. The finger service allows remote users to view the output equivalent to the **show users** [**wide**] command.
	NOTE: The previous version of the [**no**] **ip finger** command was the [**no**] **service finger** command. The [**no**] **service finger** command continues to work to maintain backward compatibility, but support for this command may be removed in some future Cisco IOS release.
Switch(config)#**no service config**	Disables the config service. The config service allowed for the autoloading of configuration files from a network server.
Switch(config)#**no ip http server**	Disables the HTTP server service

Securing End-Device Access Ports

Switch(config)#**interface range fastethernet 0/1 - 24**	Enters interface range command mode. All commands entered in this mode will be applied to all interfaces in the range.
Switch(config-if-range)#**switchport host**	Enables the switchport host macro

NOTE: The **switchport host** command is a macro that performs the following actions:

- Sets the switch port mode to access
- Enables Spanning Tree PortFast
- Disables channel grouping

The **switchport host** command does not have a **no** keyword to disable it. To return an interface to default configuration, use the global configuration command **default interface** *interface-id*:

Switch(config)#**default interface fasthethernet 0/1**

CHAPTER 8

Voice Support in Campus Switches

This chapter provides information and commands concerning the following topics:

- Attaching a Cisco IP Phone
- Verifying configuration after attaching a Cisco IP Phone
- Configuring AutoQoS: 2960/3560
- Verifying AutoQoS information: 2960/3560
- Configuring AutoQoS: 6500
- Verifying AutoQoS information: 6500

Attaching a Cisco IP Phone

Switch(config)#mls qos	Enables quality of service (QoS) globally on a switch
Switch(config)#interface fastethernet 0/14	Moves to interface configuration mode
Switch(config-if)#switchport voice vlan 210	Enables a voice VLAN on the switch port and associates a VLAN ID
Switch(config-if)#mls qos trust cos	Configures the interface to classify incoming traffic packets according to the class of service (CoS) value. For untagged packets, the default CoS value is used. The default port CoS value is 0.
Switch(config-if)#mls qos trust dscp	Configures the interface to classify incoming traffic packets according to the differentiated services code point (DSCP) value. For a non-IP packet, the packet CoS value is used if the packet is tagged. For an untagged packet, the default port CoS value is used.
Switch(config-if)#mls qos trust ip-precedence	Configures the interface to classify incoming packets according to the IP precedence value. For a non-IP packet, the packet CoS value is used if the packet is tagged. For an untagged packet, the default port CoS value is used.

`Switch(config-if)#mls qos trust device cisco-phone`	Specifies that the Cisco IP Phone is a trusted device
`Switch(config-if)#switchport priority extend cos 5`	Sets the priority of data traffic received from the IP Phone access port. The **cos** value will tell the phone to override the priority received from the PC or attached device with a CoS value of 5.
	NOTE: The CoS value is from 0 to 7, with 7 being the highest priority. The default value is CoS 0.
`Switch(config-if)#switchport priority extend trust`	Sets the priority of data traffic received from the IP Phone access port. The **trust** argument configures the IP Phone access port to trust the priority received from the PC or attached device.
`Switch(config-if)#mls qos trust extend`	Configures the trust mode of the phone
	NOTE: The **mls qos trust extend** command is only valid on the 6500 series switch. Although the 6500 series switch is not tested on the BCMSN certification exam, the **mls qos trust extend** command has been placed in this command guide because of the large number of network professionals working with the 6500 series switch.
	NOTE: With the **mls qos trust extend** command enabled, and if you set your phone to trusted mode, all the packets coming from the PC are sent untouched directly through the phone to the 6500 series switch. If you set the phone to untrusted mode, all traffic coming from the PC are re-marked with the configured CoS value before being sent to the 6500 series switch.

	NOTE: Each time that you enter the **mls qos trust extend** command, the mode is changed. If the mode was set to trusted, the result of this command would be to change the mode to untrusted. Use the **show queueing interface** command to display the current trust mode.

Verifying Configuration After Attaching a Cisco IP Phone

`Switch#show interface fasthethernet 0/2 switchport`	Displays voice parameters configured on the interface
`Switch#show mls qos interface fasthethernet 0/2`	Displays QoS parameters configured on the interface

Configuring AutoQoS: 2960/3560

TIP: QoS is globally enabled when AutoQoS is enabled on the first interface.

`Switch(config)#interface fastethernet 0/12`	Moves to interface configuration mode
`Switch(config-if)#auto qos voip cisco-phone`	Enables the trusted boundary feature. Cisco Discovery Protocol (CDP) will be used to detect the presence (or absence) of a Cisco IP Phone.
	TIP: When using the **auto qos voip cisco-phone** command, if a phone is detected, the port is configured to trust the QoS label received in any packet. If a phone is not detected, the port is set not to trust the QoS label.
`Switch(config-if)#auto qos voip trust`	Port is configured to trust the CoS label or the DSCP value received on the packet

Verifying AutoQoS Information: 2960/3560

`Switch#`**`show auto qos`**	Displays initial AutoQoS configuration
`Swithc#`**`show auto qos interface`** **`fasthethernet 0/12`**	Displays initial AutoQoS configuration for the specified port

Configuring AutoQoS: 6500

TIP: Although the 6500 series switch is not tested on the BCMSN certification exam, these commands have been placed in this command guide because of the large number of network professionals working with the 6500 series switch. The 6500 series switch uses the Catalyst operating system as opposed to the Cisco IOS found on the 2960/3560 series.

`Console> (enable)` **`set qos autoqos`**	Applies all global QoS settings to all ports on the switch
`Console> (enable)` **`set port qos 3/1 -`** **`48 autoqos trust cos`**	Applies AutoQoS to ports 3/1–48 and specifies that the ports should trust CoS markings
`Console> (enable)` **`set port qos 3/1 -`** **`48 autoqos trust dscp`**	Applies AutoQoS to ports 3/1–48 and specifies that the ports should trust DSCP markings
`Console> (enable)` **`set port qos 4/1`** **`autoqos voip ciscoipphone`**	Applies AutoQoS settings for any Cisco IP Phone on module 4, port 1
`Console> (enable)` **`set port qos 4/1`** **`autoqos voip ciscosoftphone`**	Applies AutoQoS settings for any Cisco IP SoftPhone on module 4, port 1

Verifying AutoQoS Information: 6500

`Console>` **`show port qos`**	Displays all QoS-related information
`Console>` **`show port qos 3/1`**	Displays all QoS-related information for module 3, port 1

APPENDIX

Create Your Own Journal Here

Even though I have tried to be as complete as possible in this reference guide, invariably I will have left something out that you need in your specific day-to-day activities. That is why this section is here. Use these blank lines to enter in your own notes, making this reference guide your own personalized journal.

① DETERMINE I.P. ADDRESS SCHEME.
② OVERALL DIAGRAM WITH PROGRAMMABLE ITEMS.
③ BASIC PROGRAMMING.
    a) HOSTNAME b) ADMIN VLAN1 W/ IP ADDRESS
    c) PASSWORDS d) TELNET CONFIGS.
④ L2 DEFAULT GATEWAY
⑤ ROUTER CONFIGS
⑥ L2 TRUNKING
⑦ VTP & VLAN CONFIGS.
⑧ ACCESS LINKS & VLAN ASSIGNMENT.
⑨ SVI INTERFACES & IP ADDRESSING
⑩ RE #5 - IF ROUTER ON A STICK - SVI'S DIFFERENT.
⑦a) SPANNING TREE PROGRAMMING

COMMAND SAMPLES

③ BASIC

(CONF)# HOSTNAME
    (DOMAIN)-P ENABLE SECRET
NO IP LOOKUP   LINE VTY 0 15
        PASSWORD
        LOGIN
        EXIT
        INT VLAN 1   EACH MGMT
(CONFIG-IF)# IP ADD 172.16.1.101/24 VLAN NATIVE
        NO SHUT
        EXIT.

④ L2 DEFAULT GATEWAY - IF KNOWN YET.
(CONFIG)# iP default-GATEWAY 172.16.—.1
            END

⑤ ROUTER CONFIGS (NON-ROUTER ON A STICK)
(CONFIG)# HOSTNAME___
         INT F0 or F0/0 or F0/0/0
(CONFIG-IF)# IP ADD 192.168.10.— 255.255.255.—
            NO SHUT
            EXIT.

. (CONFIG)# INT S0/0
(CONFIG-IF)# IP ADD 192.168.1.2 255.255.255.—
            NO SHUT   CLOCKRATE 56000
            EXIT
            IP ROUTE 0.0.0.0 .00.0.0 192.168.1.1 (ALL
            UNKNOWN TRAFFIC TO SERIAL)

ISP (CONFIG)# INT LOOPBACK0 (or L0)
(CONFIG-IF)# IP ADDRESS___ ___
            EXIT
(CONFIG)# ROUTER EIGRP ___
         NETWORK 172.16.1.0 (FOR EACH ADVERT.

⑥ LAYER 2&3 TRUNKING ON SWITCHES
L2
(CONFIG)# INT RANGE F0/7-8
(CONFIG-IF)# SWITCHPORT TRUNK ENCAP DOT1Q
            SWITCHPORT MODE TRUNK.
NO SHUT CHANNEL-GROUP 1 MODE DESIRABLE
            EXIT
L3.
(CONFIG)# INT F0/1
(CONFIG-IF)# NO SWITCHPORT
            IP ADD 192.168.10.6/30
            NO SHUT
            EXIT

⑦ VTP & VLAN PROG.
L2&3 SWITCHES NOT BEING SERVERS
(CONFIG)# VTP MODE CLIENT
            END.

ROUTED

SERIAL

ROUTE LOOP PROTO.

L2 TRUNK

ROUTED TRUNK

VTP SERVER PROG:
(CONFIG)# VTP DOMAIN _____
        VLAN 10
        NAME _____
        EXIT
        VLAN 20
        NAME _____
        EXIT.

(7a) SPANNING TREE VLAN ___ ROOT PRIMARY
     SPANNING TREE VLAN ___ ROOT SECONDARY

(7b) IP ROUTING. ON L3 SWITCH (SVI's)

(CONFIG)# IP ROUTING (ACTIVATE L3 ROUTING)
        INT VLAN LD
        IP ADD 172.16.___.1 /24
        NO SHUT
        EXIT
        INT.

(8) ACCESS LINKS & VLAN ASSIGNMENT.

(CONFIG)# INT F0I6
        SWITCHPORT MODE ACCESS
        SWITCHPORT ACCESS VLAN ___
        SPANNING-TREE PORTFAST
        END.

(9) HSRP SVI's. (IN PLACE OF 7b).

(CONFIG)# IP ROUTING
        INT VLAN 1
        STANDBY 1 IP 172.16.1.1
        STANDBY 1 PREEMPT
        STANDBY 1 PRIORITY 150 (#100)
        EXIT
CONFIG# INT VLAN 10
(CONFIG-IF)# IP ADD 172.16.10.3 (#172.16.10.4) 255.255255.0
        STANDBY 1 IP 172.16.10.1
        STANDBY 1 PREEMPT
        STANDBY 1 PRIORITY 150 (#100)
        NO SHUT
CONFIG# INT VLAN 20
        IP ADD 172.16.20.3 (#172.16.20.4) 255.255255.0
        STANDBY 1 IP 172.16.20.1
        STANDBY 1 PREEMPT
        STANDBY 1 PRIORITY 150 (#100)

(CONFIG)# INT VLAN 30
(CONFIG-IF)# IP ADD 172.16.30.3 (#172.16.30.4) 255.255.255.
         STANDBY 1 IP 172.16.30.1
         STANDBY 1 PREEMPT
         STANDBY 1 PRIORITY 100 (#150)
         NO SHUT.
         EXIT
(CONFIG)# INT VLAN 40
(CONFIG-IF)# IP ADD 172.16.40.3 (#172.16.40.4) 255.255.25
         STANDBY 1 PREEMPT
         STANDBY 1 PRIORITY 100 (#150)
         NOSHUT
         EXIT.

(#) means these numbers would be in plac
of the info for the second L3 switch

MAC ADDRESS of VIRTUAL LINK:
0000.0C07.AC?? - where ?? is hex
for the standby group.

(2) ROUTER ON A STICK

(CONFIG)# INT F0/0.1
(CONFIG-SUBIF)# DESC MANAGEMENT VLAN 1
         ENCAP DOT1Q 1 NATIVE
         IP ADDRESS 172.16.1.1 (24) →

         INT F0/0.100
         DESC PAYROLL VLAN 100
         ENCAP dot1q 100
         IP ADD 172.16.100.1 255.255.255.0

VERIFICATION "SHOW" CMDS

SHOW INT TRUNK
SHOW ETHERCHANNEL SUMMARY
SHOW VTP STATUS
SHOW VLAN
SHOW SPANNING TREE
SHOW IP ROUTE
SHOW STANDBY (BRIEF)
SHOW IP INTERFACE BRIEF
SHOW PORT CAPABILITIES.

_____

_____

_____

_____

_____

_____

_____

_____

_____

_____

_____

_____

_____

_____

_____

_____

_____

_____

CISCO™

CCNP Prep Center

CCNP Preparation Support from Cisco

Visit the **Cisco® CCNP® Prep Center** for tools that will help with your CCNP certification studies. Site features include:

- CCNP TV broadcasts, with experts discussing CCNP topics and answering your questions
- Study tips
- Practice questions
- Quizzes
- Discussion forums
- Job market information
- Quick learning modules

The site is free to anyone with a Cisco.com login.

Visit the **CCNP Prep Center** at **http://www.cisco.com/go/prep-ccnp** and get started on your CCNP today!